Volume 8

Panentheism

Addressing

Anthropocentrism

Man being the center of God's attention

Daniel J. Shepard

Daniel J Shepard

Channel

Panentheism Resolving the Paradox Regarding:

- ➢ Man being the center of God's attention
- ➢ God not being the center.
- ➢ The individual not being the center
- ➢ The lack of 'a' center

- Copyright 2003 – see copyright page
- Panentheism, anthropocentrism, metaphysics, human purpose, title
- Second Edition (July 4, 2014)
- Second Printing (July 4, 2014
- ISBN-13: 978-1503020757
- ISBN-10: 1503020754
- CreateSpace.com
- W.E.Hope, Inc. (World Embracing Hope)
- 1. Panentheism 2. Geocentrism 3. Panentheism 4. Symbiosis 5. Determinism 6. Title

Panentheism
Addressing
Anthropocentrism

A gift

From me to you

From one soul to another

Peace

The Gift

Copyright:

Any part of this trilogy may be reproduced or utilized in any form or by any means. The formal copyright was obtained only to protect the source and the integrity of the work and to guarantee your access and authorization to freely use and reproduce this work. These concepts are not my own. They are the merging and logical conclusion to the blending of ideas created within a society supported and maintained by vast numbers of people, including you and those who came before you.

Daniel J Shepard
Channel

Panentheism
Addressing
Anthropocentrism

Note to the reader:

- The intent of the more than 20+ books is to provide enough material to prove the validity of panentheism not beyond 'all' doubt but to prove the validity of panentheism beyond 'all reasonable' doubt. The point being to elevate individual's and our species' perception of themselves in order to elevate human behavior on both an individual level and on a species level before we begin to step into the heavens.

- The series of books, Panentheism, emerged from earlier metaphysical editions and have been edited and retitled to more accurately reflect the true nature of their contents.

- I understand there are numerous stylistic, grammatical and spelling errors within all my work. I hope you as a reader can overlook such issues and focus upon the ideas being presented. I do not like to make excuses but all the material is, after all, free to the public and therefore producing no revenue stream.

  Having spent more than a quarter of a million dollars on the web site: panentheism.com, 20+ books, presentations, videos, attempts to place the material in the hands of academics and the public ... I found my resources insufficient for formal editing. It is perhaps best to consider the products of my work more as a personal log in the rough of what it is I have been entrusted, with the condition that I pass this material on to you.

Daniel J Shepard
Channel
Panentheism.com

Daniel J Shepard
Channel

Panentheism
Addressing
Anthropocentrism

Books by Daniel J. Shepard

<u>Panentheism</u>

Vol. 1:   Panentheism addressing Humanity's Purpose
Vol. 2:   Panentheism addressing Man made in the Image of God
Vol. 3:   Panentheism addressing Sci./Rel./Phil./and Prophecy
Vol. 4:   Panentheism addressing Volumes 1 – 3 Guide
Vol. 5:   Panentheism addressing The Physical and the non-Physical
Vol. 6:   Panentheism addressing Humanity Confined to a Universe
Vol. 7:   Panentheism addressing Free Will and Determinism
Vol. 8:   Panentheism addressing Anthropocentrism
Vol. 9:   Panentheism addressing Theodicy
Vol. 10: Panentheism addressing Ethics
Vol. 11: Panentheism addressing the Lack of 1st Cause
Vol. 12: Panentheism addressing $E = mc^2$
Vol. 13: Panentheism addressing The Mathematics of non-Members
Vol. 14: Panentheism addressing Creation/the Void
Vol. 15: Panentheism addressing Monism/Dualism
Vol. 16: Panentheism addressing Nihilism
Vol. 17: Panentheism addressing Language
Vol. 18: Panentheism addressing Philosophy's Responsibility
Vol. 19: Panentheism addressing Ockham's Razor
Vol. 20: Panentheism addressing Panentheism
Vol. 21: Panentheism addressing being 'the' Summit
Vol. 22: Panentheism addressing History's Vector
Vol. 23: Panentheism addressing Western Philosophy
Vol. 24: Panentheism addressing Chaos/Complexity
Vol. 25: Panentheism addressing Abbreviated Thoughts
Vol. 26: Panentheism addressing The Whole of Reality
Vol. 27: Panentheism addressing The Soul
Vol. 28: Panentheism addressing God/Brahma
…

More information can be found at my web site

❖

www.panentheism.com

Daniel J Shepard
Channel

The intent of the more than 20+ books is to provide enough material to prove the validity of panentheism not beyond 'all' doubt but to prove the validity of panentheism beyond 'all reasonable' doubt.

The point being to elevate the individual's and our species' perception of themselves in order to elevate human behavior on both an individual level and on a species level before we begin to step into the heavens.

In today's environment it appears faith could use the assistance of rationality to overcome the forces of skepticism, relativism and nihilism. It is the intent of the series, Panentheism, to provide just such assistance.

Panentheism
Addressing
Anthropocentrism

Panentheism, a small seed planted into the social fabric of our species. An idea which only takes one Greek word to express, 'panentheism' and three English words to explain, 'pan' all, 'en' in, 'theism' God. 'All in God' and with that simple phrase our species has the potential to change forever.

Author

Project Overview

1995 - 1996 Final draft of "You and I Together: Have a purpose in reality" completed. This was a process of coalescing forty years of thoughts regarding a Universal Holistic System. From these notes, a model was constructed. The impact was then examined regarding this particular model and the effect it would have upon humanity in terms of the most cherished concepts embraced by the individual as well as those embraced by our species.

1996 - 1997 Final draft of "In the Image of God" completed. This step involved testing the practicality of a Universal Holistic System. The work examines the ability of the System to resolve twenty futuristic socially-divisive issues and ten current socially-divisive issues.

1997 - 1998 Final draft of "Stepping up to the Creator" completed. Once the system had been developed, the impact examined, and the practicality tested, the Universal System needed to be formalized, expanded, and validated against what it is we believe - religion, what it is we observe - science, what is we reason - philosophy, and what it is we've been told about change - prophecy. The work takes on a three-dimensional matrix format. The matrix format was used to help the reader move in and out of the 900 various topics and levels of difficulty.

1998 - 1999 Final draft of the Cross Reference Guide and Index" completed. Because of the expansiveness of the project, the need arose to find a means of cross-referencing the intricacies of the project. This was accomplished through the development of a cross-reference sectioned into five categories: Questions Addressed, Flowcharts, Thematic Index, Index, and Glossary.

1999 First draft of CD completed: The project was converted into Adobe Acrobat format. This was done to make the project user-friendly. The CD assists the exploration of the project through the power of the search engine called Adobe Acrobat. The CD will be updated as the project progresses.

1999 First draft presently unfolding on site of "On 'being' being 'Being'" This is a technical work intended for deep thinkers. Its intent is, through constructive criticism, to examine the error of humanity's perceptual journey generated by philosophers over the last twenty-five hundred years. The Universal Holistic System of Panentheism acts as the foundation of the constructive criticism.

Daniel J Shepard
Channel

1999 First draft of CD completed: Multimedia presentation of the www.wehope.com project as well as other misc. lectures. This series of lectures/presentations is made in person. Even philosophers must strive to apply practical applications to their work. The W.E. Hope Foundation is a nonprofit organization established by this philosopher in an attempt to apply the fundamental principles he espouses.

1999 CD - Part I. Audio readings of articles. The CD's are custom made. Please link to www.wehope.com for additional information.

1999 CD - Part II. Audio readings of articles. The CD's are custom made. Please link to www.wehope.com for additional information.

2000 Multimedia Presentation - A Universal Philosophy. This is a 981-slide presentation, in Adobe Acrobat format, that explores the means by which we could attain a universal philosophy. This presentation will be available for online viewing later this year.

2000 In the articles section of the Library page, a number of articles are available for viewing. These are works-in-progress and are intended to be incorporated into a new trilogy to be completed later this year.

2000 A new page "Reflections" has been added to the site. These are an account of my thoughts and reflections on a variety of philosophical issues and questions.

2000 A new page "Aphorisms" has been added to the site.

2000 A new page "Definitions" has been added to the site.

2000 - 2003 The final tractate of the third volume of a new trilogy was placed online. The complete trilogy - The War & Peace of a New Metaphysical Perception - introduces a new perceptual model of reality. The work is intent upon establishing the understanding of a new metaphysical system, which combines the Aristotelian metaphysical system of Cartesianism and the Hegelian metaphysical system of non-Cartesianism into one system. The three volumes of the new trilogy are as follows:

    2001: Volume I - On 'being'

    2001: Volume II - On 'being' being

    2001: Volume III - On 'being' being 'Being'

2003 – 2005 Existence: In and of Itself - Introductory Work to Trilogy II: The War and Peace of a New Metaphysical Perception.

Panentheism
Addressing
Anthropocentrism

2004 Convert and place on line: The War and Peace of a New Metaphysical Perception to an Ontological Version.

2004 – 2005 Convert complete site from HTML to CSS / DHTML to stabilize site for the long term and to facilitate removing and reinstalling site if it becomes corrupted through use or hacking.

2005 New Site Appearance, Complete All Sections of the site except 'Latest Additions', Add additional sections to the site, and Complete Final Appearance of Site.

2005 - 2008 Move the project to the more advanced interactive www tool of blogging: Adding reason to faith URL: http://panentheism.blogharbor.com/

2009 Development of a new series: Understanding ...

2010 Understanding Reality: The four absolute truths secularists are intent upon eradicating are: 1. A Creator of the physical universe exists. 2. The true essence of the individual is made in the image of this Creator and is thus, by definition, divine in nature 3. The individual and our species exist temporarily in the physical for a reason. We have a purpose. 4. The void, ex-nihilo, creation from non-existence did occur. These four fundamental, absolute truths will be addressed in great detail within this book and will, beyond all reasonable doubt, be shown to exist as absolute truths. The theists need more than faith to establish their positions in this day and age and this work gives them what they need to rationalize their positions.

2011 Converting the work into a format compatible to createspace.com and kindle.com. Placement of work onto createspace.com and kindle.com.

2012 Panentheism: Addressing the Whole of Reality

2013 Understanding the Soul

2015 Understanding God/Brahma

Daniel J Shepard
Channel

Panentheism
Addressing
Anthropocentrism

Panentheism

… is the only understanding of reality rationally capable of addressing …

Anthropocentrism

Man being the center of God's attention

Daniel J. Shepard

Daniel J Shepard
Channel

To Err Is
Human
To Forgive
Divine

Alexander Pope

The error: The paradox of Centrism

Panentheism is the only understanding of reality rationally capable of validating human purpose as individuals and as a species while removing the concept of anthropocentrism.

The perception: Copernicus moves our perceptual understanding regarding the system being filled with Centrism into that of being 'the' system filled with Centrism and non-Centrism. As such, Centrism and non-Centrism, with the help of Copernicus, now have a location within which they can be found. However, the understanding regarding the role of both Centrism and non-Centrism as well as the understanding regarding the interrelationship between Centrism and non-Centrism not only remain in a state of confusion but even more disconcerting, the existence of such an interrelationship is not recognized as a significant aspect of the 'larger' system.

It is this state of this confusion which will be specifically addressed within this tractate.

Panentheism
Addressing
Anthropocentrism

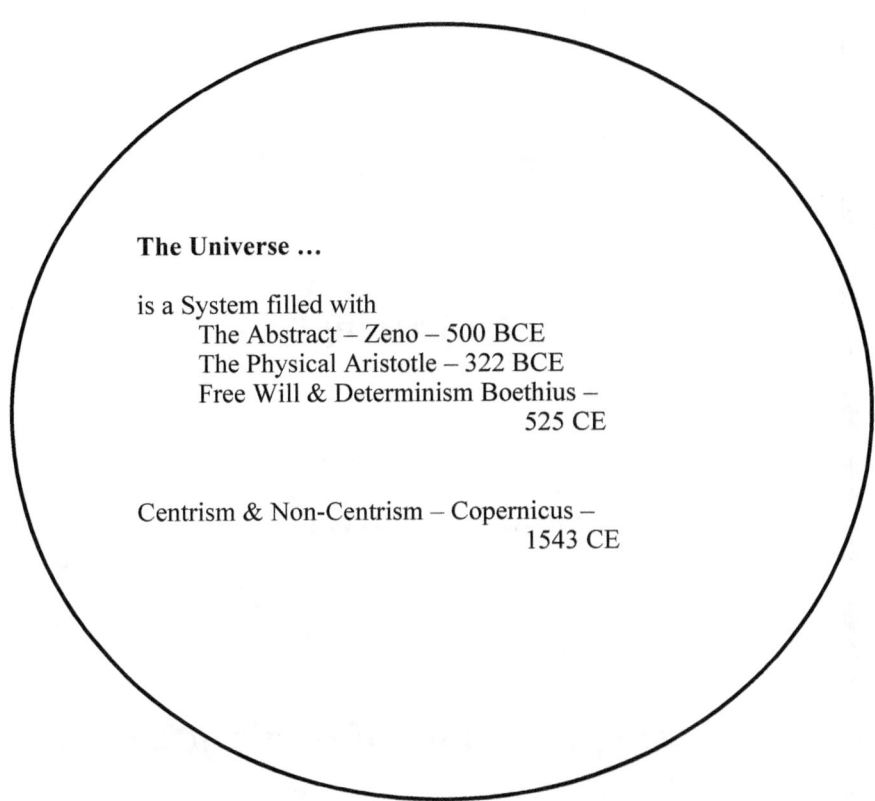

The Universe ...

is a System filled with
  The Abstract – Zeno – 500 BCE
  The Physical Aristotle – 322 BCE
  Free Will & Determinism Boethius –
                                525 CE

Centrism & Non-Centrism – Copernicus –
                                1543 CE

Understanding Evolving[1]

---

[1] For the year 2000 CE, Humanity's entry into the 3rd millennium see page 163

Volume 8

Panentheism

Addressing

Anthropocentrism

Man being the center of God's attention

•

Daniel J. Shepard

Channel

Panentheism
Addressing
Anthropocentrism

Table of Contents   Page(s)

Part I: The Paradox of Centrism and non-Centrism   21

1. Introduction I   21
2. Introduction II   23
3. Pre-Copernican   25
4. Post-Copernican   27
5. Copernicus' paradoxes   31

Part II: Resolving the paradox of Centrism with a new metaphysical perception   35

6. Centrism   35
7. A location of Centrism   37
8. A location of non-Centrism   39
9. The dynamics of Centrism   41
10. The dynamics of non-Centrism   47
11. The law of inverse proportionality   57
12. The 'location' of 'nothingness'   67
13. Virgin physicality/'virgin physical life'   87
14. Virgin consciousness/'virgin abstract knowing'   91
15. Stepping 'in' beyond Centrism: Dependency   95
16. Stepping 'into' Centrism: Independence   103
17. The significance of insignificance: Random Sequencing   113
18. The explosive nature of the potentiality of knowing   127
19. Removing a piece of Randomness   129
20. Boethius' metaphysical system and why we can now file it away as a part of the annals of history   131
21. Archimedean Points   139
22. Philosophical infinities   141
23. A bag of marbles is not dependent upon sequential time   147
24. A unit of knowing is not a marble   151
25. Symbiotic Panentheism   163
26. About the author and the work   165

Daniel J Shepard
Channel

Panentheism
Addressing
Anthropocentrism

Terms/concepts

'Fundamental building block' of the abstract
'Fundamental building block' of the physical
'Inverse proportionality'
'Virgin physicality/'virgin physical life'
Aristotelian Points
Centrism
Hegel's 'open' dynamic non-Cartesian system
Kant's 'closed' dynamic Cartesian system
Non-Centrism

Daniel J Shepard
Channel

Panentheism
Addressing
Anthropocentrism

Anthropocentrism

- Man being the center of God's attention
- God not being the center.
- The individual not being the center
- The lack of 'a' center

Part I: Creating the paradox of a Centrist System

1. Introduction

With the scientific understanding of the sun being the Cartesian[i] point of origin, the point (0,0,0) as opposed to the individual being the Cartesian point of origin, the individual was literally put into motion.

By moving the concept of the center from being a man, to the center being the sun, Copernicus in essence created the perception of humanity having lost its sense of being the center. Such a perceptual development introduced two concepts into our understanding of the universe.

First: we now perceived the universe to have a center, scientifically speaking. Second: Philosophically speaking, we now perceived the universe to be the whole within which 'a' center could be 'found'.

Before Copernicus' revelations, we had philosophically perceived 'the center' to be that of 'knowing'. Pre-Copernicus, we perceived the human id not only 'representing' the center of reality but also 'being' the center of reality.

With Copernicus' observations, humanity lost its concept of being the center and as such, philosophy/reason found itself being moved from the forefront to being place behind science/observation, and eventually being removed from second place in line to being placed third in line.

This second transition of moving from second place to third place occurred with the cultural elevation of God/religion via Christianity, Islam, etc to second place.

How does Copernicus' development of Centrism and non-Centrism differ from the Aristotelian development of Cartesian and non-Cartesian?

Daniel J Shepard

Channel

The Aristotelian system of Cartesian and non-Cartesian emerged as a static physical system while the Copernican concepts of Centrism and non-Centrism moved the perceived Aristotelian 'static' system into a perceived 'dynamic' system of movement.

With the acceptance of the Copernican system, everything became objects in motion and all motion took place around a center, thus Centrism.

In fact, with the acceptance of Copernican system, everything became objects of motion centered around the Aristotelian point of origin.

The Aristotelian system led to the Copernican system, which was to lead to the Kantian system.

This historical evolution was to open up the concept of system to Kant's dynamic 'closed' Cartesian system. Kant's system would in turn open up the concept of 'systems' and allow for the development of Hegel's suggestion of 'the' system being a dynamic 'open' non-Cartesian system.

The development of both Kant's and Hegel's systems in turn created the potential acceptance of a new metaphysical model which was to follow their innovations.

The new system model which was to follow Kant's 'closed' dynamic Cartesian system and Hegel's 'open' dynamic non-Cartesian system was a model perceived as a 'dynamic open non-Cartesian system powered by a dynamic closed Cartesian systems' or better labeled as the individual 'acting within'/being a part of God or more generically speaking, panentheism.

However, we are moving too rapidly, therefore let's begin again:

Panentheism
Addressing
Anthropocentrism

2. Introduction II

Copernicus was not a philosopher but Copernican perceptions immensely influenced philosophical thought.

The perception initiated by Copernicus: The system is a location for Centrism as well as for non-Centrism.

This perception moved Aristotle's Cartesian perception from that of being a static system to that of being a dynamic system and in essence introduced the concept of Centrism co-existing with non-Centrism within our universe, within our reality.

As we found with Aristotle, scientifically introducing two extremes of a concept may assist the workings of science but it can greatly befuddle our abstractual perceptions of philosophy and religion.

To unravel the philosophical and religious riddle introduced through the elevation of the significance of Centrism to that of non-Centrism, we must first understand the pre and post scientific perceptions existing before Copernicus' Centrism was established as a scientific 'fact'.

Daniel J Shepard
Channel

Panentheism
Addressing
Anthropocentrism

3. Pre-Copernican

Before Copernicus, the center of existence resided with the individual. One might even suggest that pre-Copernicus, the center of existence resided 'within' the individual.

The individual was the center of 'knowing' since the individual was where 'knowing' resided.

This is not implying that gods or God were not recognized entities.

Rather it suggests that gods and God were perceived to be humanistic in form or at the very least capable of presenting themselves as such.

Pre-Copernicus, 'knowing' was the center of existence. Zeno acknowledged this with his suggestion that although abstraction might be a separate entity found 'within' the universe, it existed as a separate entity nonetheless.

Philosophy, religion and science all reinforced each other in terms of the center.

Philosophy viewed the individual as the center of reason, the center of knowing. Religion viewed the individual as the center of attention, the arena around which the gods centered their attention.

Science viewed the individual as the center of all that exists, the heavenly bodies all revolved around the individual, revolved around 'knowing', revolved around awareness.[ii]

Now it was understood that men did move from place to place. However, it was also presumed by groups of men, which the heavenly bodies revolved around their particular group, revolved around their particular location of 'home'.

Daniel J Shepard
Channel

Panentheism
Addressing
Anthropocentrism

4. Post-Copernican

With Copernicus, the center of physical reality began to move outward from the individual.

Science/ observation began to seek the center and as the process of seeking the center evolved, the center was found to move 'outward', move away from humankind.

The understanding regarding 'where' the center was 'located', moved from being 'within' the individual to being… well, we no longer knew where the center was to be found. It was the center we now sought to find.

With Copernicus, the sun became the center of the concrete/physical. In spite of the fact that the sun now becomes the center, the concept of 'the center' remains and as such, the universe eventually evolves into being the concept with a center via the big bang.

Confusion over the centrist perception not only remains in terms of the concrete but the concept of Centrism remains in terms of perceptions regarding abstractual concepts. Humankind 'looks to' the center of all things: the center of the concrete and the center of all abstraction.

Scientifically the center was 'probably' 'out' there somewhere.

The quest for the scientific Holy Grail became the quest to find the center, the quest to find the primary origin of both the universe and life.

The more science looked, the further removed the center became from the individual.

In the macroscopic sense the center moved from the center of 'knowing' found within the individual to the sun, to the center of the galaxy, to the center of the universe.

In the microscopic sense the center moved from the center of 'knowing' found within the individual to the cell, to the nucleus of the cell, to the nucleus of the atom, to the quark, to…

As the search for the center moved away from abstractual 'knowing' and into the physical, science took off as 'the' source of knowledge.

Daniel J Shepard

Channel

As the reputation of science being the legitimate tool for finding the 'center' increased, the legitimacy of philosophy and religion diminished.

The further the center became removed from the individual the more insignificant the individual became.

Insignificance was not necessarily increasing in terms of human behavior but insignificance was increasing in terms of humanity's own perception regarding the rationality of human significance.

As time progressed, tolerance and respect due the individual was increasing but tolerance and respect were not increasing due to the increase in the rational understanding regarding why such respect should exist but rather tolerance and respect were increasing based upon the argument: We should tolerate and respect each other because...

'Because why?' was the question and the answers centered on answers such as: Because we say we should.

Because we believe we should. Because that is the way I want to be treated. Because God said we should. Because...[iii]

Where were the answers involving the rationality of such behavior? The further we moved the center away from ourselves, the further removed the rationale regarding tolerance, respect, human compassion; abstract hedonism became removed from our understanding these very aspects of human knowing.

Our significance became simply a grain of sand in the beach of time as the center of origination moved further and further outward from ourselves.

As the center moved further and further away from ourselves, we lost the understanding regarding our significance for our significance became lost in time, space, and perceptual understandings regarding the limitlessness of reality versus Reality.

Reality, with an upper case 'r' became lost and as such religion and philosophy became confused, humankind became lost and confused.

Energy – matter are dualities of the physical universe. Time – distance are dualities of the abstractual universe. Both dualities are dualities of our personal universe.

Are they the only forms of physical existence or abstractual existence found within our personal universe?

We are not naive enough to believe, with a fair amount of certainty, that this is not the case regarding abstractual concepts found within our personal universe.

Panentheism
Addressing
Anthropocentrism

We are naïve enough to believe, with a fair amount of certainty, that is the case regarding physical concepts found within our personal universe.

Why are energy and matter something in which we 'believe' regarding the physical but time and distance are not something in which we 'believe' regarding the abstract?

We believe the physical to be composed of only matter and energy because that is all we can observe/measure at this point in time.

We believe the abstract is composed of more than time and distance because there are more than these two ideas of which we are consciously awareness regarding 'knowing' itself.

One may say, however: We can measure time and distance.

That is true, and so it is we may find time and distance to be aspects of the physical, to be innate characteristics of the physical[iv], to be aspects unique to the physical rather than aspects of abstractual existence itself.

That too, however, is another topic.

Are we going to postpone our discussion of such a topic as we have so many others?

Yes and no, for we have already touched upon this very idea in both tractates one and two and we will delve into the concept of time and distance within practically every tractate found within Book II of this trilogy.

For the time being, however, we must stay on task and examine the concept of Centrism and non-Centrism, which Copernicus has so eloquently placed before us.

Daniel J Shepard
Channel

Panentheism
Addressing
Anthropocentrism

5. Copernicus' paradoxes

First a person's home, then city, country, continent, planet, sun, and galaxy 'had a center', was 'the' center. Now it is the origination of the Big Bang, which is the center.

The origination of the Big Bang is now the point of origination, the center, we seek to 'find'.

The 'primal atom', the point from which our universe began its expansion is what we desperately seek to find as we literally turn our earth into a massive radio telescope.

It is the 'center' we seek to find as we send huge telescopes into orbit around our little inconspicuous planet.

The concept of a center, the concept of 'a' point of origination, haunts us because we perceive time to be the one and only true existence. We have no perception of the concept regarding a 'location' of timelessness.

The only way to eradicate or change the perception we have of ourselves is to change our perception.

To be truly a change, such a perceptual shift must metamorphose our present understanding regarding our immense insignificance into becoming an understanding of our phenomenal significance.

Such a task is no easy matter.

Such a task cannot emerge from science/observation; rather such a metamorphosis must emerge from philosophy/reason and in particular from metaphysics.

It is observation of the physical/science whose function it is to aid us in understanding the degree of our insignificance.

It is reason/philosophy, metaphysics in particular, whose function it is to aid us in the understanding the degree of our significance.

Why is this the case? This is the case because science deals with the physical and philosophy deals with the abstract.

Again, we see the validity of Zeno's assertion that both the physical and the abstract exist as independent entities dependent upon one another.

Where then does our third means of developing perceptions enter the picture? Religion, our ability to believe, our third means of developing perceptions, finds its function lies in aiding us in believing what it is we observe and what it is we find reasonable.

This is not to say that everything we observe or reason is 'fact'.

Rather it simply says: If we cannot believe in anything we observe or believe in anything we reason, then there is nothing but religion itself left in which we can believe.

We presently have a fairly secure scientific/observational understanding regarding who we are and how it is that we, humans, awareness, a packet of knowledge, an entity of abstractual consciousness, function as an entity of abstractual consciousness existing in a region where we are immersed in time rather than time being immersed within ourselves.

Since we basically understand the concept that our physical essence lies within time and space, we will not be examining such a concept within the limits of this tractate.

We presently, however, do not understand who we are and how it could be that we, humans, awareness, a packet of knowledge, an entity of abstractual consciousness, could function as an entity of abstractual consciousness existing within a region where we are not immersed in time but rather time is immersed within us.

The understanding of this concept will be the focus of our attention in Part II of this tractate.

What do the two previous paragraphs have to do with Copernicus, Centrism, and non-Centrism?

The first 'We presently...' paragraph deals with Copernicus' initiation of Centrism being the principle found within our universe, found within our reality.

It is this concept, which fundamentally defines Zeno's concepts of 'multiplicity' discussed in Volume 5.

The second 'We presently...' paragraph deals with Copernicus' inadvertent initiation of potential non-Centrism being the principle found 'outside' the universe, found 'outside' our reality, found 'within' the greater Reality.

It is this concept, which fundamentally defines Zeno's concepts of 'seamlessness' discussed in Volume 5.

Panentheism
Addressing
Anthropocentrism

How do we begin to understand such an alien concept as time existing 'within' us when we exist in a region of timelessness?

How does such a concept even begin to relate to Copernicus himself?

The whole concept of understanding such seemingly non-understandable concepts had its foundation laid within Tractates one, two, and three. Zeno, Aristotle, and Boethius all influenced our way of thinking and as such subconsciously directed us to a conclusion regarding Copernicus' revelations, conclusions Copernicus may have also deduced but by no means stipulated in his observations.

Copernican astronomy: The system of astronomy that was proposed by the Polish astronomer, Nicolas Copernicus (1473 –1543) in his book, *De revolutionibus orbium coelestium*, which was published in the month of his death and first seen by him on his deathbed.

It used some elements of 'Ptolemaic astronomy', but rejected the notion, then current, that the earth was a stationary body at the center of the universe.

Instead Copernicus proposed the apparently unlikely concept that the sun was at the center of the universe and that the earth was hurtling through space in a circular orbit about it.[v]

What was it Copernicus suggested that had such a great influence upon our philosophical train of thought?

Copernicus moved the concept of 'center' beyond our reach and it has remained beyond our reach ever since.

The concept of center moved off our planet, moved beyond our ability to 'go there'.

As our abilities to 'go there' expanded, the 'center' of our 'known' reality moved further and further from our reach.

One thing that did not change, however, was our perception that a center to 'it all' could be found and so it is we look for the origination of 'Om' the sound emanating from the center of origination itself, the sound originating from the center of the universe itself, the sound originating from the center of reality itself.

Since Copernicus, we have clung to the concept of Centrism and rejected the very idea of a non-centrist existence.

So it is, abstraction and the physical remain as Zeno's seamlessness and Zeno's multiplicity confined 'within' reality.

Daniel J Shepard
Channel

So it is we ignore the concept of a greater 'Reality'.

So it is the Aristotelian concept of the physical and the abstract being 'contained' within the lesser reality of the physical universe maintains its status of authenticity while the potential existence of a greater Reality, a reality with no center maintains its status as an occult form of metaphysical existence.

How sad it is that we have allowed the once proud intellectual arena of metaphysics to fall to such a lowly status.

How sad that we have wrapped ourselves within the boundaries of the physical, only to feel its wrappings shrink in upon our psyche and sense our sanity slipping into a form of mimicry Gerard so aptly describes.

We must find the means of cutting the wrappings confining our psyche.

We must emerge from the imprisonment of our own making or we shall surely collapse what small sense of abstractual hedonism still exists and we shall find ourselves completely seduced by the charms of physical hedonism.

Panentheism
Addressing
Anthropocentrism

Part II: Resolving the paradox of Centrism with a new metaphysical perception

6. Centrism

Philosophically we have been immersed within the concept of Centrism since time began.

1. We have seen ourselves as the center of concern.
2. We have seen our tribes as the center of sociological development
3. We have seen 'Om' as the center of the universe.
4. We have seen the Earth as the center of the universe.
5. We have seen the Sun as the center of the solar system.
6. We have seen the nucleus as the center of the atom.
7. We have seen God as the center of creation.
8. We have seen ourselves forever searching for the center, for the origination point of the universe.

Philosophically and religiously, Centrism has not disappeared just because Copernicus scientifically demonstrated that the earth is not the center of the solar system.

We still seek the center.

We are not only scientifically, but philosophically and religiously enamored with the concept of 'a' 'center'.

We understand there is a location for multiplicity.

We now understand there may well be 'a' separate location for the 'lack of multiplicity'/seamlessness. (Volume 5: Zeno)

We have now seen there is a location for the physical. We have seen there may well be 'a' separate location for abstraction. (Volume 7: Aristotle)

We have observed there is a location for free will. We have observed there may well be 'a' separate location for determinism. (Volume 7: Boethius)

Daniel J Shepard

Channel

By the end of this tractate, we will be able to make a fourth statement. We will be able to state:

We recognize there is a location for Centrism. We recognize there may well be 'a' separate location for non-Centrism. (Volume 8: Copernicus)

The rationality regarding our universe, our reality, being a part of a larger 'Reality', grows with each tractate.

Panentheism
Addressing
Anthropocentrism

7. A location of Centrism

With Zeno, Aristotle, and Boethius, we understand the characteristics of the universe to be:

> The Universe
> Physical reality
> The concrete
> The abstract
> Time/distance

Time and distance exist 'within' the physical because time and distance are innate characteristics of matter and energy[vi].

We can add the abstractual understanding regarding the concept of Centrism to this graphic since Copernicus elevated the concept of Centrism to the level of being a basic principle of science for many centuries following his observations.

The graphic may therefore be expanded as:

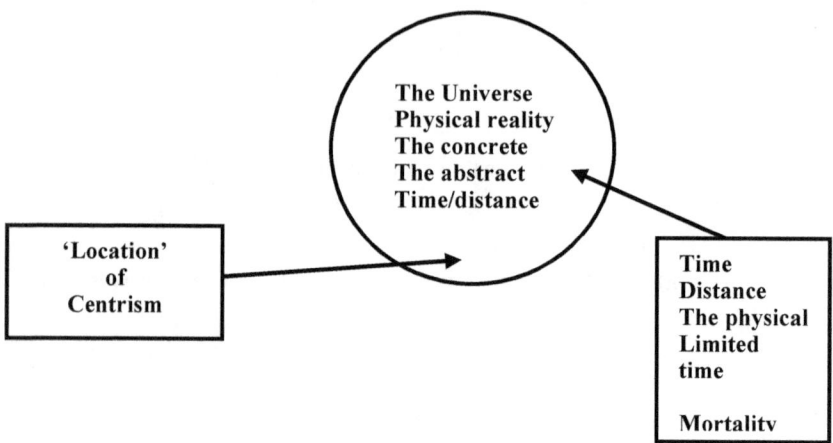

Rather than spend large quantities of time building a rational for the existence of an 'outside' to the universe, let's simply add an outside to the universe.

To better understand the rationale for such a leap, one needs to retrace their steps and read Tractates 1, 2, and 3.

With this said we can move immediately into the next section and add the 'outside' to the graphic.

Panentheism
Addressing
Anthropocentrism

8. A location of non-Centrism

The location of non-Centrism:

The characteristics:

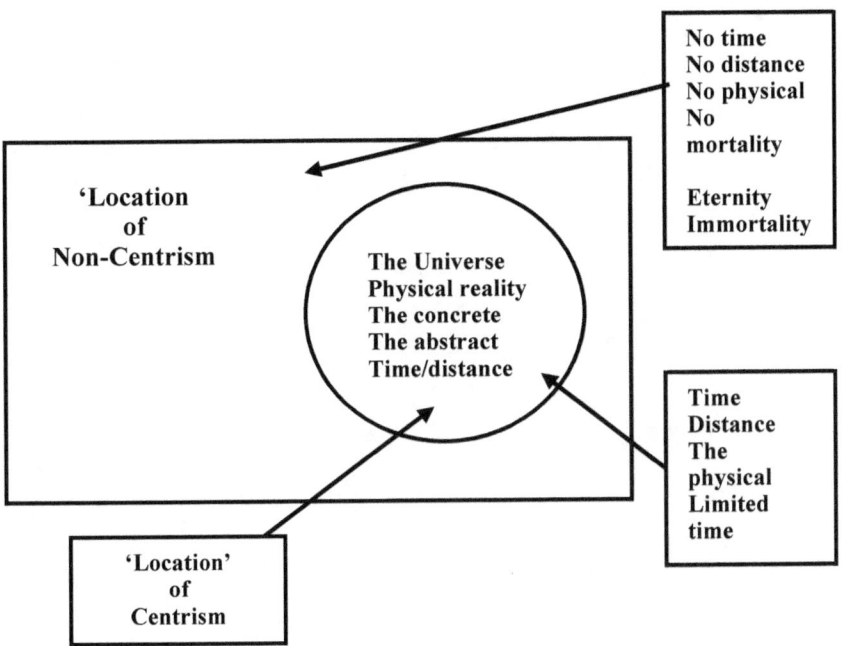

Time and distance exist 'within' the physical because time and distance are innate characteristics of matter and energy.

Time and distance do not exist 'outside' the 'location' of the physical since time and distance are innate characteristics of the physical and the physical cannot be found 'outside' the physical itself

Daniel J Shepard
Channel

Panentheism
Addressing
Anthropocentrism

9. The dynamics of Centrism

Centrism involves two primary concepts: time and distance. Without time and distance no center can be found.

Time and distance invoke the concept of motion, finding, seeking, looking for, reaching for, ... the point of origination, the point (0,0,0), ...

To understand the dynamics of Centrism, let's remove the location where Centrism is not the dominating concept:

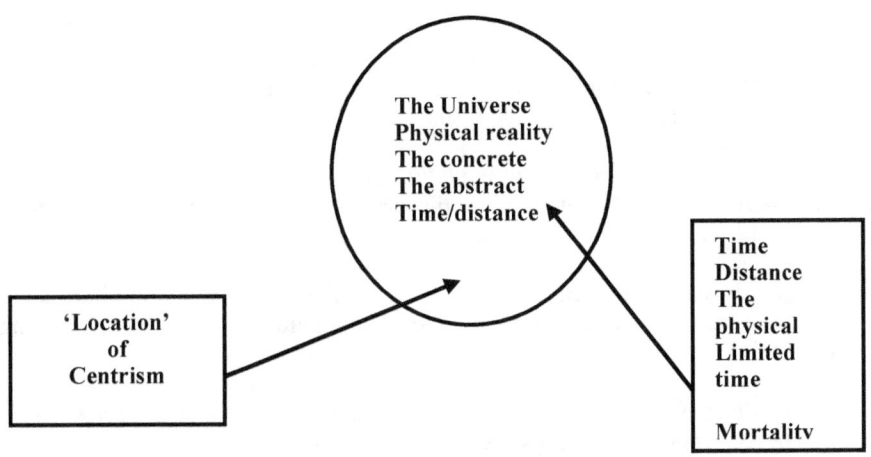

We understand the dynamics of the universe to be what the dynamics of the universe are as perceived by ourselves and what we, at this point in time, perceive the universe to be is a region permeated with the concept of beginnings leading to endings.

We understand the initial step in the process is the concept of 'beginning' for without 'a' 'beginning' we cannot rationalize an 'ending' to what it was which we were focusing our thinking process or meditation upon.

Not only do we perceive the universe to be filled with concepts of beginning – end but also science perceives the universe itself to be affected by such linear events as beginning – end.

In fact, many scientists are looking for the very process of beginning – end, which they believe, may apply to the universe itself.

Other scientists, unable to understand the answer to the question,

'What existed before the universe?' are attempting to understand the mechanism required which would explain the concept of the universe having 'no' beginning and 'no' ending.

The major hurdle confronting scientists today in terms of building a workable model of a 'timeless' universe is time itself.

Time appears to permeate our universe from one end to the other.

Time in fact appears to be an innate characteristic of matter and energy themselves.

It was Einstein who suggested: $E = mc(2)$. 'c' being the velocity of light in a vacuum is simply d/t and thus not only are matter and energy implied to be directly dependent one upon the other but also matter and energy are demonstrated to be absolutely related to both space/distance and time.

As such, as long as we perceive the model of the universe to be a model depicting the universe to be 'filled' with matter and energy, we will perceive the universe to be 'filled' with time.

Does this imply we must discard our perception of the universe existing in order to resolve the paradox of Centrist systems, resolve the paradox of all 'thing's having a point of origination?

We cannot discard our perception of the universe existing and expect such an action to be 'acceptable' to beings 'existing' within such a location.

We can however build a metaphysical model, which leaves the universe temporarily intact and functioning for trillions of years or more.

We can build a metaphysical model, which demonstrates a function for such an existence and as such, a function for the entities found 'within' this subset of the whole, found 'within' the element we call 'our' 'reality', the universe.

How does one accomplish such abstractual tasks? One takes the first step and accepts the characteristics of time, and thus the characteristics of matter, energy, and space, to be infinite within its own characteristics of time itself but finite 'because' it is just that: time.

Panentheism
Addressing
Anthropocentrism

At first glance, this may appear a strange action to take, but the precedent of accepting apparent dual contradictory characteristics is an idea that has already been set by science.

It was science that first suggested we accept the characteristics of a photon to include characteristics of both matter and energy.

Since science established the concept of potentially accepting dual contradictory characteristics, there is no reason we cannot apply the concept of duality to time, space, matter, and energy.

In the metaphysical model suggested of an 'outside' to the physical, we are applying the duality of infiniteness and finiteness to the concept of this new metaphysical perception.

Time is infinite for it has no end until the end occurs and time is finite in nature for time is limited to the concept of time itself.

Many would object to such a contradictory statement regarding time. Many would argue: There is no way such a perception can be 'proved'.

That is the same argument humanity used to explain why man cannot fly, why man cannot get to the moon, etc.

To continue our examination of a new metaphysical perception we are going to do what individuals such as the Wright brothers and Van Braun did.

We are going to ignore critics and explore the uncharted.

We are going to explore the concept that time is infinite by means of its linear characteristic but finite by means of its own existence.

Having made the decision to accept the duality of finiteness and infiniteness of time, we will accept the same of matter, energy and distance/space.

But why accept the same of three new concepts, the concepts of matter, energy, and distance/space?

We place the same parameters upon matter, energy, and distance/space as we do upon time because science directly ties the four together through the mathematical language of science via the equation $E = mc(2)$.

Daniel J Shepard

Channel

When science distances itself from their own basic perception and replaces the perception of E = mc(2) with a new perception expressed mathematically and accepted by the scientific community universally then we will metaphysically have to reexamine what it is that we are about to do in establishing a new metaphysical perception.

With this having been said we are now ready to explore the dynamics of a non-centrist location. We will do so by first establishing such a location.

So again, we come back to …

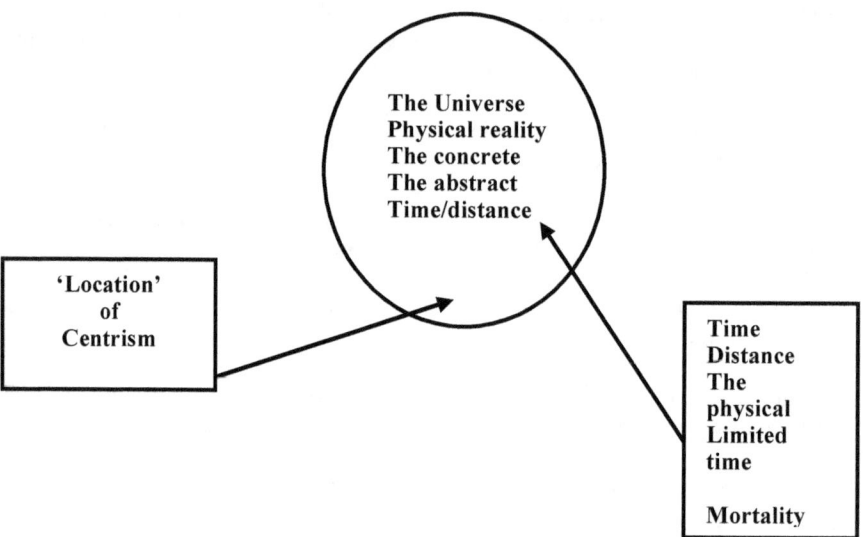

… in order to build our location of non-Centrism.

Panentheism
Addressing
Anthropocentrism

We build the location of non-Centrism through the simple action of designating a location of non-Centrism:

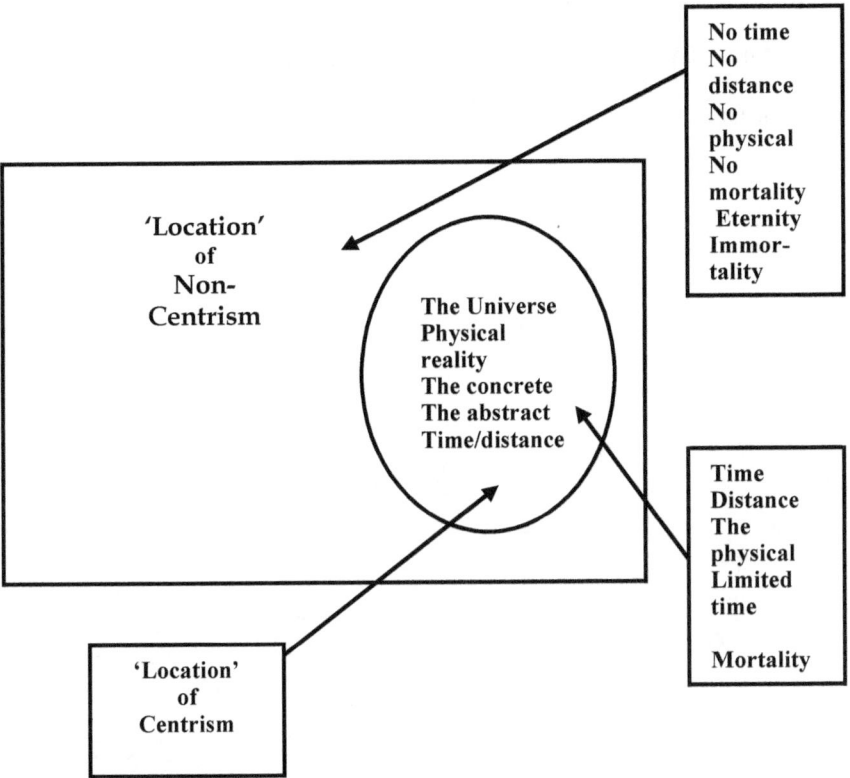

The first question, which arises, is: Why place the location of non-Centrism 'outside' the physical, outside the universe, outside 'reality'?

We are not going to address that issue in detail in this tractate since it was examined in detail in Tractates 1 and 2.

It should be stated at this point, however, that the diagram is not intended to suggest that Centrism can only be found within the universe and therefore cannot be found 'outside' the universe.

Nor does the diagram suggest that non-Centrism can only be found 'outside' the universe and therefore cannot be found 'within' the universe.

What the diagram suggests is that it is Centrism, which is the predominant characteristic found 'within' the universe and that it is non-Centrism, which is the predominant characteristic, found 'within' the region 'outside' the universe.

Having said this and having established the location of non-Centrism, we can now begin an examination of the dynamics regarding the 'location' we call a location for non-Centrism.

Panentheism
Addressing
Anthropocentrism

10. The dynamics of non-Centrism

A region of non-Centrism lacks four primary elements found within a region where Centrism dominates.

Non-Centrism lacks the pervasive presence of time, space/distance, matter, and energy.

Ignoring concepts of matter and energy for now, we can begin an examination of abstractual concepts of time and distance.

Extracting the concepts of matter and energy from the region of Centrism will allow us to examine the dynamics of time and distance/space.

The concept of a beginning, an origination, the origin, loses its meaning within a region we call non-Centrism.

In a region where matter and energy do not exist as a part of the universal fabric, time and distance/space are elements, which in turn do not exist as part of the universal fabric.

Without the perceived abstractual concepts of time and distance/space, the concept that a center can be found loses its validity as a fundamental principle of existence and thus the understanding of the concept of a non-centrist location begin to emerge.

How is it possible to find the lack of time and distance within the universe?[vii]

The only way to find such a 'location' within the physical is to look to the non-physical, look to abstraction.

Is this to say the 'location' of non-Centrism is found within the physical? No, rather it is to say we have within our reality the indication that a 'location' of non-Centrism does exist.

We have, within our physical reality, non-centrist concepts which we can observe but which we cannot measure physically.

We have, within our physical reality, non-centrist concepts we believe but which defy faith.

We have, within our physical reality, non-centrist concepts we find logical but which reach beyond our rationality.

Daniel J Shepard

Channel

In spite of our primitive understandings regarding abstractions, we refuse to acknowledge the possibility of there existing 'a' location where abstractions can exist 'outside' the physical.

Our refusal to acknowledge the possible existence of 'a' location where abstractions exist outside the physical partially emerges from the implications such an existence implies.

Time and distance are concepts that provide the means by which we understand what it is to move from 'here' to 'there'.

Zeno described such movement as incremental elements of multiplicity.

Zeno, however, did not ignore the abstractual. Zeno, through his introduction regarding the paradox of space and time, introduced the concept of seamlessness.

Seamlessness, the lack of time and the lack of distance, invokes the scientific concept regarding the lack of motion.

Without 'a distance' through which to move, motion becomes a perceived contradiction and time becomes an unnecessary element.

Without the existence of time and distance/space, the concept of motion becomes irrelevant.

The concept of motion in and of itself creates the need for time within which to move from point B to point C.

Thus without distance/space and time, motion becomes irrelevant.

The lack of time and distance does not, however, become a problem just for science.

A lack of a point of origination for motion, a lack of a beginning, in essence, a lack of creation becomes what at first glance might be perceived as a fatal blow for religion. In fact, however, it does no such thing.

Philosophy, in particular metaphysical models, in a region void time and space/distance, experiences similar apparent setbacks, as do science and religion.

Without time, philosophically and metaphysically, the problem of a beginning as initiated by a linear progression of time would appear to suggest there is no need to find the source of knowledge, no need to seek 1st truth, no need to look for 1st cause, no need to reach for the meaning of life itself.

Panentheism
Addressing
Anthropocentrism

Movement involving relocation through the process of incremental multiplicity, which the physical incorporates, need not be one of relocation from one physical location to another physical location.

There are two forms of movement.

There is physical movement, which is characterized by the principle of multiplicity and there is abstractual movement, movement characterized by the principle of seamlessness.

Abstractual movement can be found in the 'location of Centrism' and might better be characterized as simply a process of thought displacement.

Although abstractual movement, thought displacement, seamlessness of motion, can be found within a location of Centrism, such movement would best describe the primary form of motion found within the 'location' of non-Centrism.

To best understand such concepts let's examine the concept of physical motion existing within the location of Centrism.

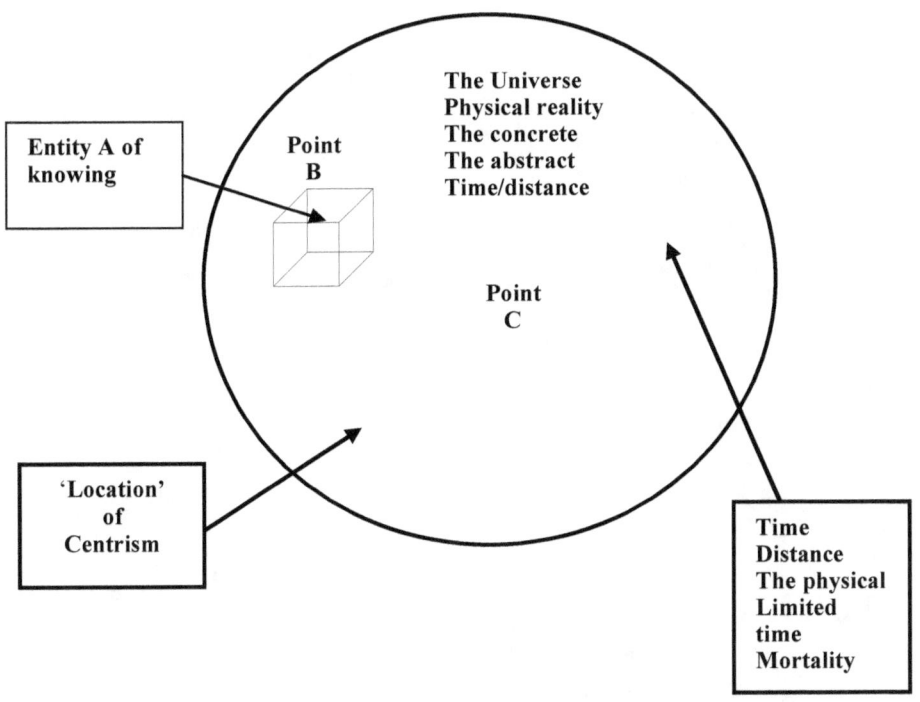

For the entity of 'knowing - A' to move from point B to point C, 'A' must traverse distance/space, which takes the function of time to accomplish since distance/space is a characteristic of matter/energy, which produce the innate characteristic of time itself.

This process of 'physical movement' is what could be explained as an orderly process of sequential actions or simply sequential orderliness.

Little more need be said regarding such movement since we are quite familiar with this type of motion, which surrounds us on a daily basis.

This brings us to examining 'motion' within' the location of non-Centrism. In fact, such an examination leads us directly to understanding just why such a location is called non-Centrism.

Panentheism
Addressing
Anthropocentrism

To understand movement within non-Centrism, we must rebuild the location of non-Centrism.

To do so we will take the latter graphic and expand it through the inclusion of the location of non-Centrism:

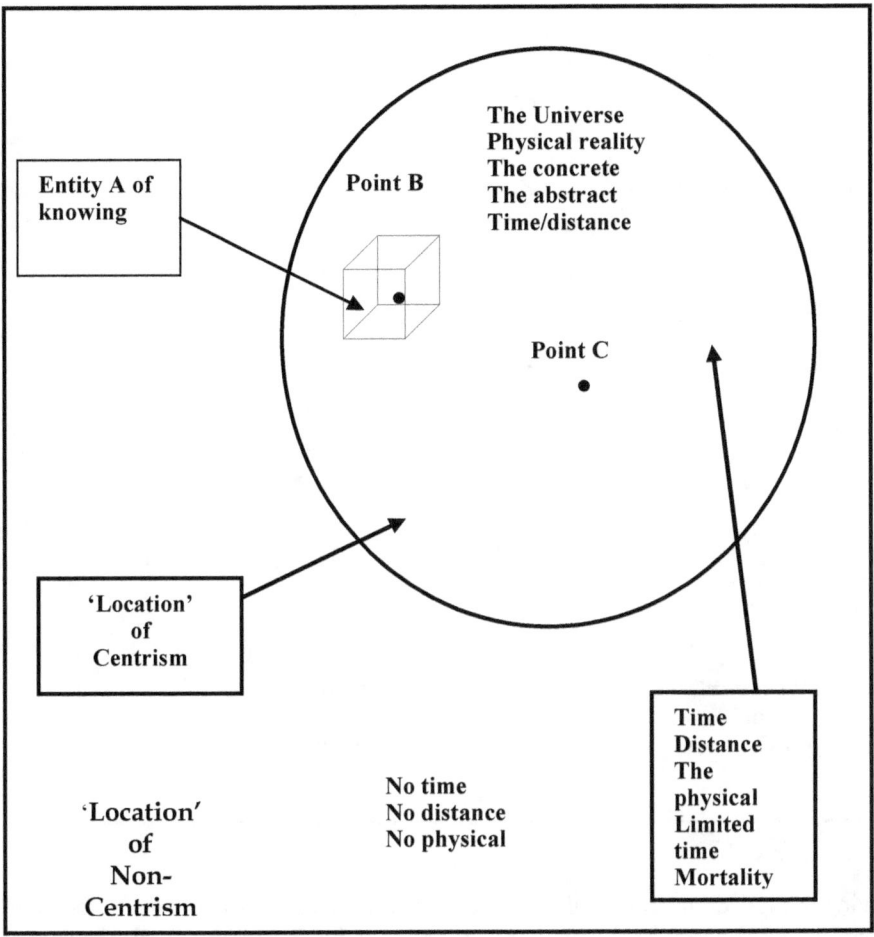

To simplify the discussion we will extract the region we have already explored. If we remove the physical, we obtain:

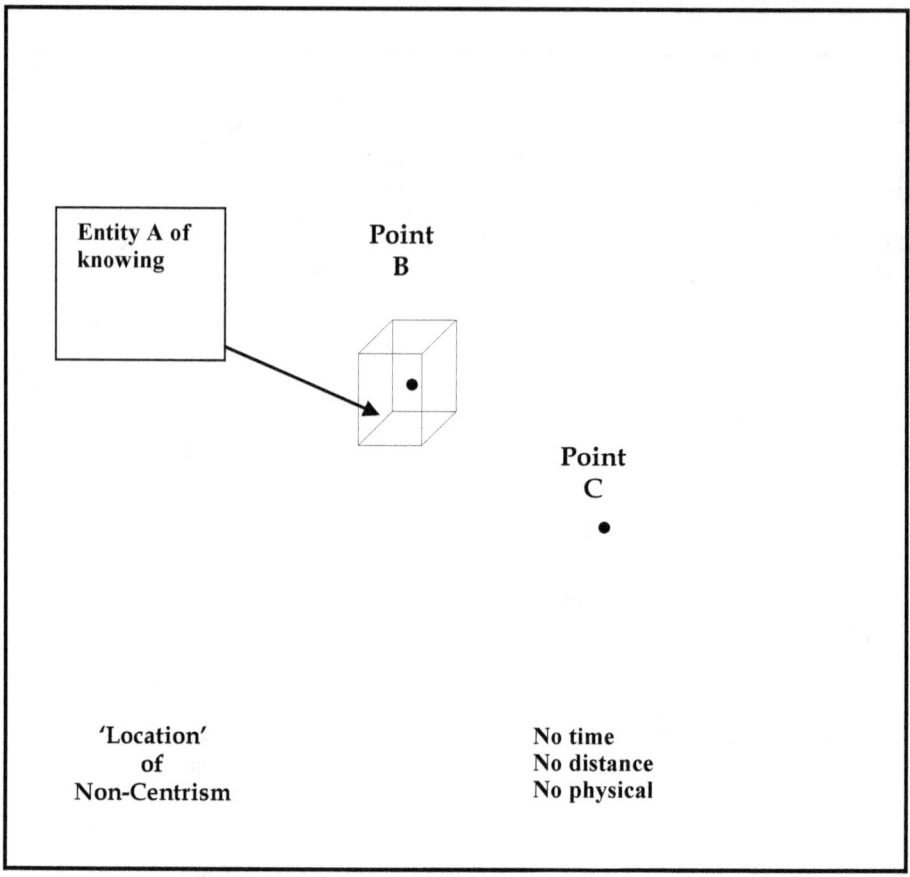

Having removed the physical, time, and distance/space, we obtain a location where points B and C still exist but to go from point B to Point C takes no time since there is no physical distance between them.

Panentheism
Addressing
Anthropocentrism

The ramifications of such a perception may better be understood in terms of the interaction, which occurs between several entities of knowing.

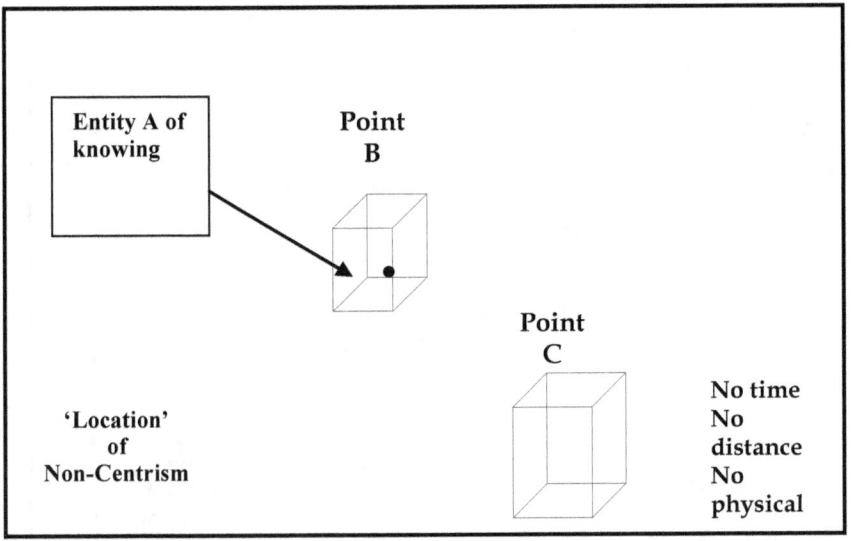

In essence, this depiction now becomes:

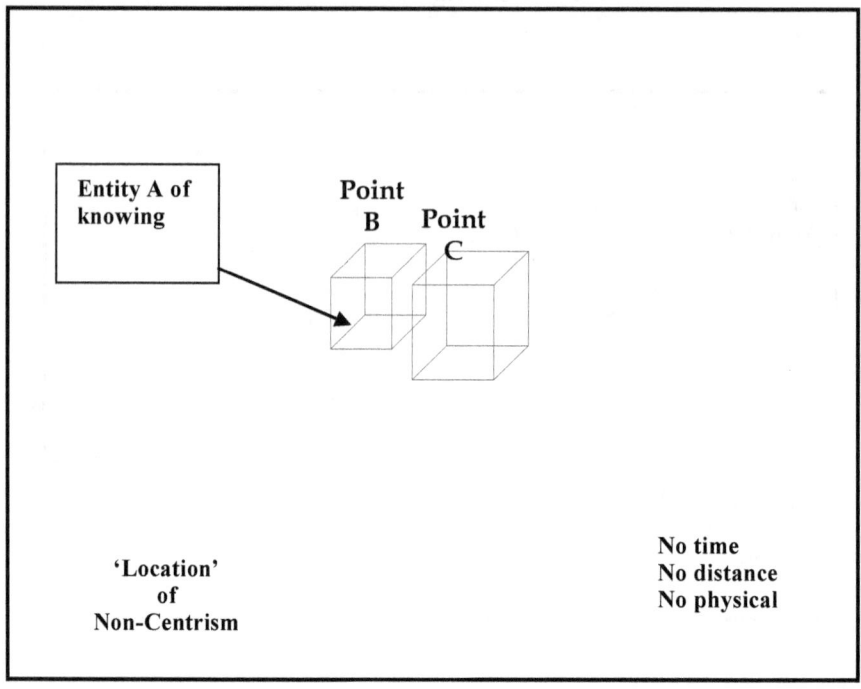

The 'distance' between point B and C no longer exists 'between' points B and C.

Distance/space and time do not disappear.

Distance and time remain but remain as abstractual items of knowing found within the knowing of each entity itself.

Each entity of knowing, which experienced time and space through the process of 'traveling' through time and space, traveling through the physical, traveling through reality, traveling through the universe, has retained its awareness and understanding of just such experiencing.

Panentheism
Addressing
Anthropocentrism

Lets examine the graphic with many entities of knowing:

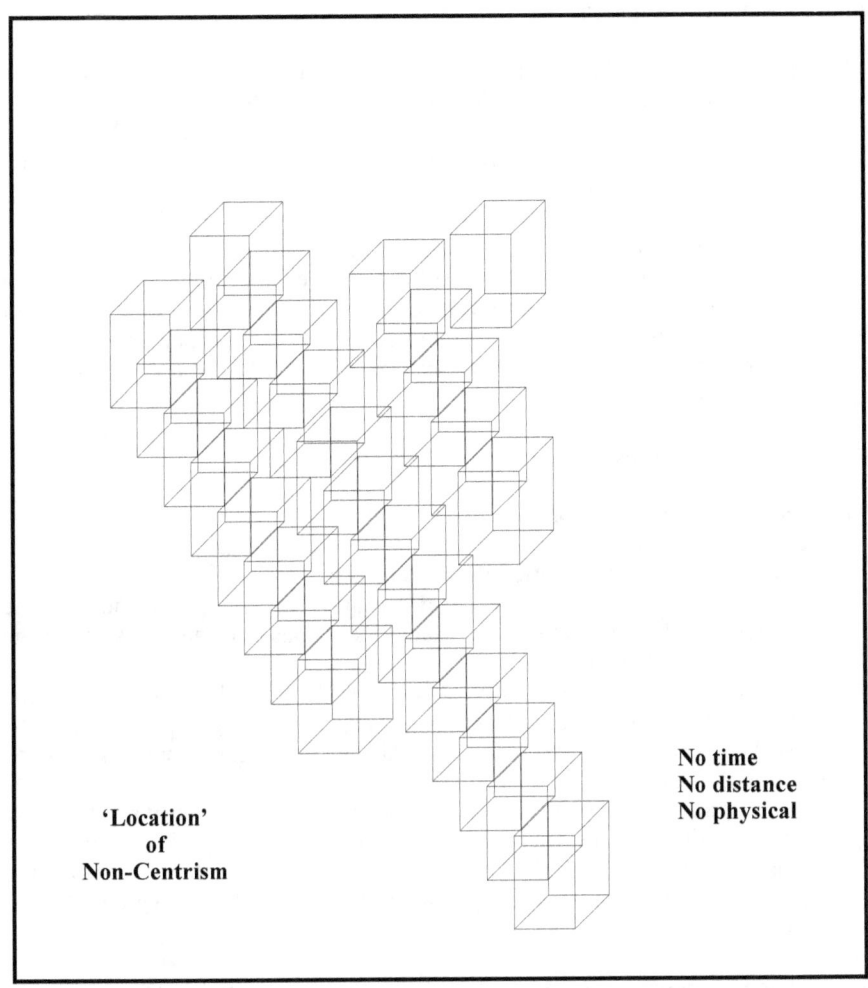

'Location' of Non-Centrism

No time
No distance
No physical

Such a perception initiates an understanding regarding what might be called a form of Brownian movement of thought, of abstraction.

Brownian abstractual principles may in fact, apply equally to abstractual concepts found within a region of non-Centrism as it applies to physical concepts found within a region of Centrism.

Metaphysical perceptions now begin to emerge as the foundation of free will and the individual.

Within a location of Centrism:

1. Sequential orderliness of thought, sequential orderliness of cause and effect, emerges as a universal characteristic of Centrism
2. One can impact 'what is' and what 'will be' or what 'could be' since 'what is' is too short a time to exist and 'what will be' or 'what could be' do not yet exist.
3. One develops as a unit of unique knowing through the action of free will which eventually adds to the location of non-Centrism
4. One retains one's own unique self but can expand upon what it is one understands with the incorporation of the understanding of one's own self in relationship to what exists around one's self. This expansion is limited in scope by the limits of what exists as opposed to the non-existence of 'what will be'.

Within a location of non-Centrism:

1. Non-sequential randomness of thought, non-sequential randomness of cause and effect, emerges as an expansion of perceptual possibilities found as a universal characteristic of non-Centrism
2. One can impact 'what was' and 'what is'. Experiences such as the existence of Hitler's actions and Gandhi's actions exist and thus create the personality of the whole itself.[viii]
3. One develops the potentiality of the non-Centrism location through the introduction of one's unique knowing acquired through 'present' actions invoked while in the realm of space/distance and time found within the realm of Centrism.
4. One retains one's own unique self but can expand upon what it is one understands with the incorporation of unique units of knowing which exists around one's self. This expansion, is unlimited in scope by the infinite potential of newness 'what will be' affords the region of non-Centrism.

These statements incorporate the concept of orderliness and randomness.

In the Western scientifically oriented society, the term 'orderliness' is viewed as a positive characteristic and the term randomness is viewed as a negative characteristic.

Within this tractate, the concepts of positive and negative characteristics are not the issues being addressed. Rather we are examining the very basics of existence itself.

Panentheism
Addressing
Anthropocentrism

11. The law of inverse proportionality

Science speaks of the principle of symmetry:

For every action there is an equal and opposite reaction, positives and negatives, up and down, matter and anti-matter, energy and? (anti-energy?).

Religion initiated this principle with its concept of 'good' and 'evil'.

But what of philosophy, what is it philosophy has put into place regarding the principle of symmetry?

It might be stated that in terms of symmetry, philosophy has put forward the concept of life and death.

Such a perception initiates the concept of death being a form of 'evil', death being a negative as opposed to the positive of life.

If we were to graph such perceptions, we would obtain:

Scientifically:

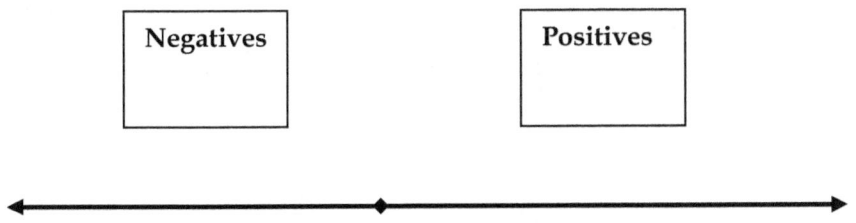

If one then combines the positive with the negatives one obtains:

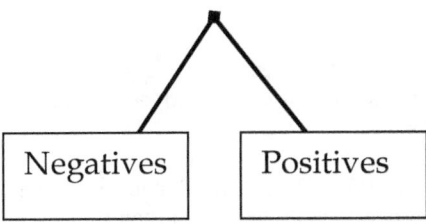

And then

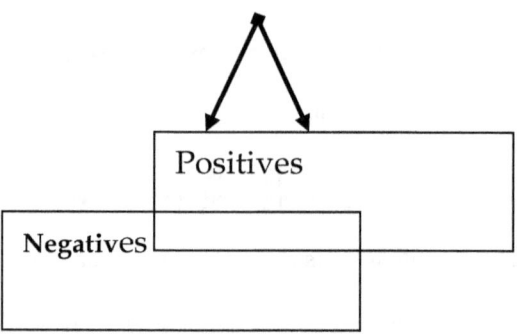

And finally:

▼

Or:

O

'Nothing'

The same process can be applied to other scientific perceptions: For every action there is an equal and opposite reaction, up and down, matter and anti-matter, energy and ? (anti-energy?).

Applying the principle of symmetry to 'good' and 'evil' becomes the central aspect of many religious perceptions regarding the 'eternal' conflict of 'good' and 'evil'.

Panentheism
Addressing
Anthropocentrism

Philosophically if we apply the same concept of symmetry to 'life' and 'death' we obtain:

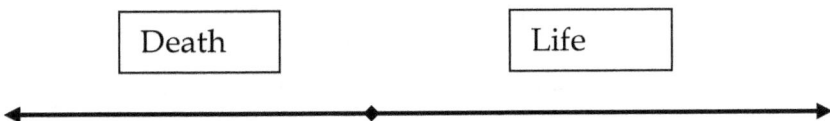

This makes no rational sense in terms of our present day perceptions regarding 'death'. Scientifically death means 'end'.

What then has philosophy to offer our understanding regarding the perception of life and death in terms of the principle of symmetry?

Philosophy has nothing to offer our understanding regarding life and death but confusion and uncertainty.

This confusion and uncertainty spills over to religion and science.

The perception diagramed, having no rationality associated with it, causes the very principle of symmetry to take on the perception of being a faulty principle.

The perception regarding the principle of symmetry being flawed, attaches a sense of doubt to the scientific and religious principles involving symmetrical perceptions.

If one philosophically accepts death for what it is, one obtains:

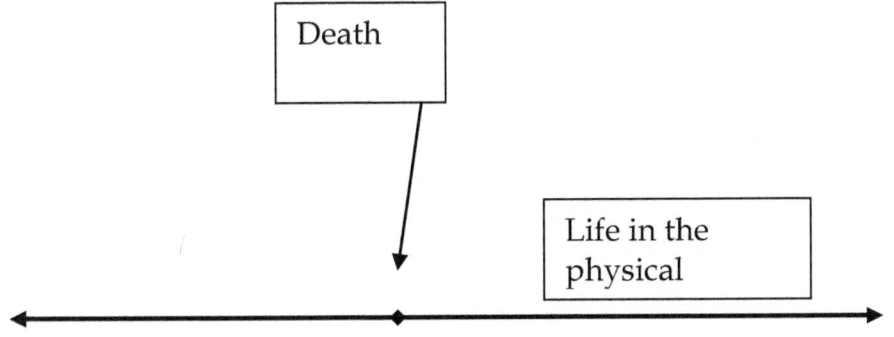

And then:

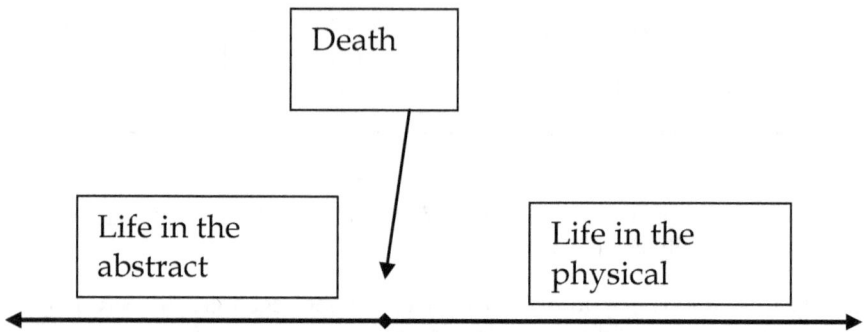

This perception moves 'death' from being the opposite of 'life' to being a 'cross-over' point as opposed to a form of existence.

Why does such a graph suggest that death becomes a 'cross-over point' as opposed to a form of existence itself? Death has no past and no future.

Death is an existence in the present.

If we apply the same graphic actions to this philosophical depiction as we did to scientific depictions we obtain:

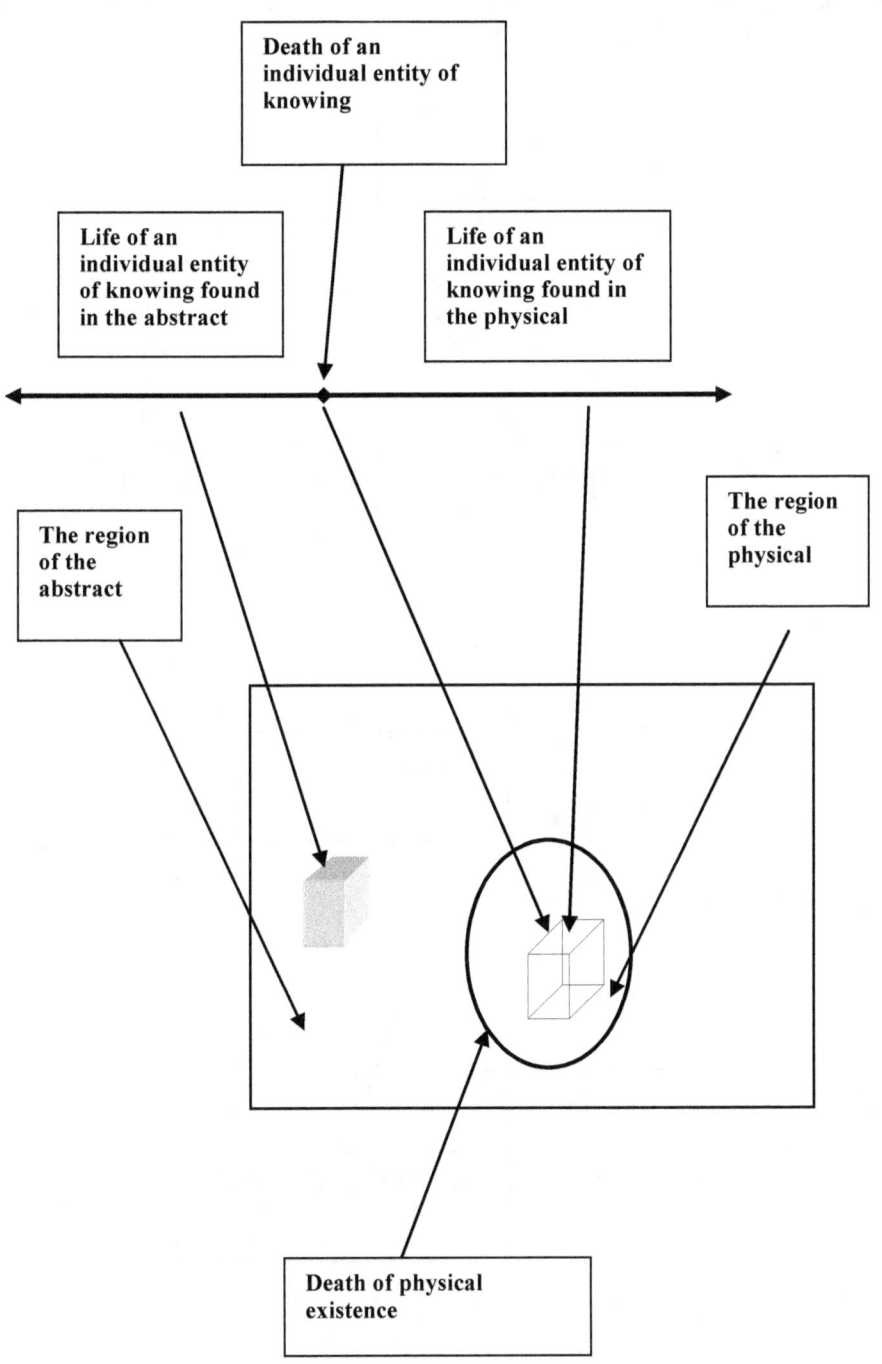

Which, applying the graphics demonstrating the scientific summation of the concepts positive and negative, we obtain:

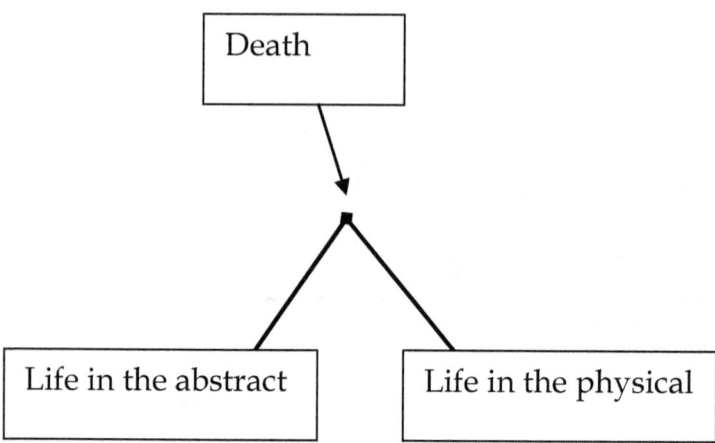

And then

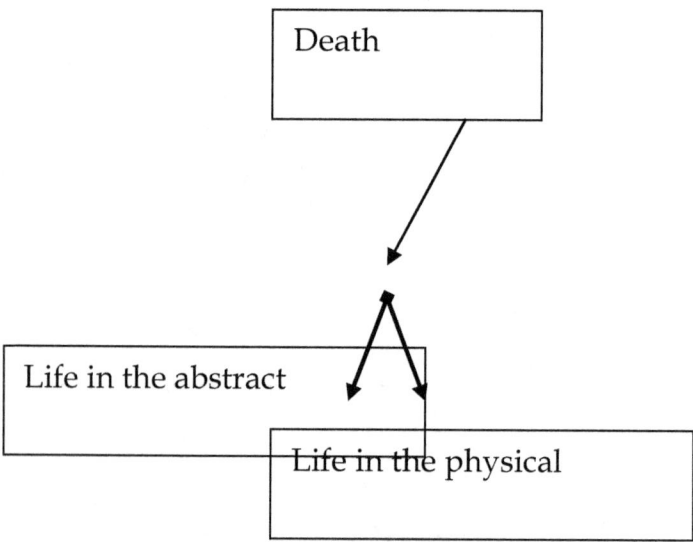

Panentheism
Addressing
Anthropocentrism

And finally ...

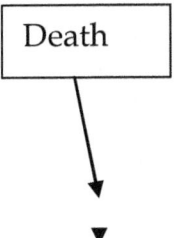

... or once again ...

Death

O

... 'nothing'

Such a philosophical perception may seem insignificant until one realizes it brings philosophy into line with science and religion.

It is philosophy, which takes a lesson from religion/believing and science/observing means of developing perceptions.

Is such a lesson important for philosophy?

Absolutely.

Philosophy has been 'stuck' long enough in terms of resolving its most prominent paradoxes.

It is time for philosophy to resolve its most perplexing puzzles and move on.

Philosophy has no reason to fear it's becoming an outmoded form of perception.

Daniel J Shepard

Channel

Resolving the most intriguing philosophical paradoxes will not bring an end to philosophy.

Other paradoxes are awaiting discovery but these other paradoxes must wait for philosophy to accomplish its present task, the task of understanding the relationship between the abstract and the physical.

Once accomplishing this task, philosophy will find that new, exciting, and even more challenging frontiers await.

Such an event occurred with science and it will occur with philosophy.

How can philosophy accomplish its desire to resolve the philosophical paradoxes confronting it for the last twenty-five hundred years?

Philosophy simply needs to expand its concept of 'reality' to that of 'Reality'.

Philosophy simply needs to embrace the physical with the abstract:

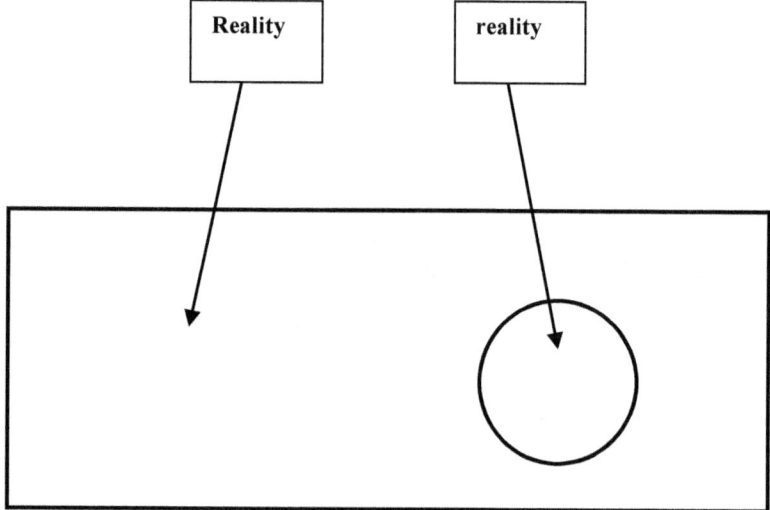

Panentheism
Addressing
Anthropocentrism

And place the individual entity of 'knowing' where we know it to be, within the physical ...

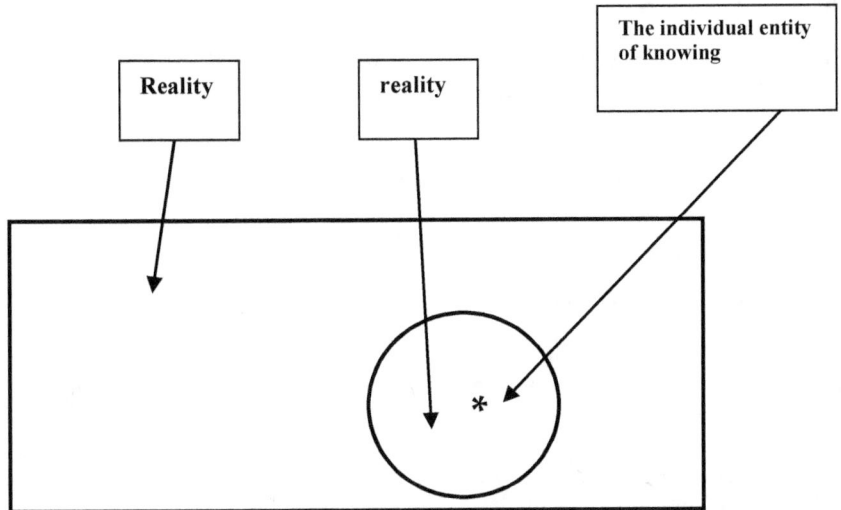

... which is the metaphysical system of the individual acting within God, panentheism.

With the application of the principle of symmetry to philosophy, we find the paradoxes of philosophy become manageable and allow philosophy to move on with its task of providing science and religion new perceptions and paradoxes intended to challenge their creative thought and development.

This process does not appear to be a form of 'inverse proportionality'. This segment of the Tractate appears to deal with the law of symmetry. Therefore why title it: 'The law of inverse proportionality'?

Philosophy has an obligation to advance thought into active forms of imagery rather than cement thought into static forms of imagery.

The law of symmetry, as we saw demonstrated, suggests all concepts, be they physical or abstract, can be reduced to 'nothingness' itself.

This is not what the progressive diagrams illustrated within this section are implying.

What is being implied is something new.

Daniel J Shepard

Channel

What is being implied is the existence of an active form of imagery.

What is being suggested is that the abstract exists within the location of abstraction and the abstract also exists within the location of the physical.

One location becomes the 'real' and the other becomes a 'real illusion' yet at the same time, the 'real illusion' becomes the 'real' and the 'real' becomes the 'real illusion'.

Which is which depends upon one's point of reference.

This concept was fully addressed within Tractates 1, 2, & 3.

As such, we will not explore such a discussion other than to say:

Philosophically speaking, 'all' does not reduce to 'nothingness', rather 'nothingness is the mirror separating one side from the other, separating the 'real' from the 'real illusion'.

As such, the principle of symmetry rather than being the principle of symmetry is suggested by metaphysics to be the law of inverse proportionality.

This is no insignificant statement.

This statement could revolutionize scientific and religious perceptions.

And who said philosophy had nothing new to offer our perceptual development?

But why use the analogy of nothingness representing a mirror as opposed to nothingness being a location of reductionism where all reduces to the lack of everything including the lack of nothingness itself.

A mirror is functional. It is not a glass separating one from the other.

If one views oneself in a mirror and spends some time examining the opposing image, one realizes that the image is not the same as the object.

With some detailed observation, the person viewing their image realizes that their right hand is the images left hand.

Thus, a transformation takes place between the object and the emergence of the object's image as projected by the mirror.

In essence, a form of inversion of the object becomes the image and thus becomes the law of inverse proportionality.
This proportion will become an important element in the two tractates, Volume 11: Hegel and Volume 13: Einstein.

Panentheism
Addressing
Anthropocentrism

12. The 'location' of 'nothingness

The left becomes the right and the right becomes the left at the boundary.

Where is this boundary?

The boundary is best described as a mirror.

When looking into the mirror one will notice that as one lifts their right hand the mirror image lifts its left hand.

As one scratches one's right ear the mirror image scratches its left ear.

It is often quoted that the eye is the mirror into the soul of a man.

What then of the eye one views when looking into the mirror?

Is it one's own soul or the soul of the whole, the soul of God Itself into which one peers?

And what mirror is it of which we speak when speaking of 'nothingness itself'.

It is nothingness, which separates the 'Real' from the 'real'. It is nothingness, which acts as the boundary separating the 'real' from the 'real illusion' (Volume 5: Zeno).

It is nothingness, which acts as the boundary separating the Cartesian from the non-Cartesian (Volume 7: Aristotle).

It is nothingness, which acts as the boundary separating 'free will' from 'determinism' (Volume 7: Boethius).

And now we in essence are exploring the concept of nothingness separating Centrism from non-Centrism (Volume 8: Copernicus).

What do we perceive as 'nothingness' has its location as the mirror itself.

Nothingness is and nothingness functions as the zero point on a number line.

Daniel J Shepard

Channel

Nothingness allows one side to 'see' the other side, to see the image as we saw in the Tractates dealing with the concepts Boethius, Aristotle, and Zeno placed before us.

Each philosopher had a dilemma with which to deal and each dealt with their particular philosophical dilemma in their own unique manner.

Zeno, Aristotle, Boethius, and now Copernicus could not see the mirror just as we cannot see the mirror as we gaze into the mirror itself.

They each 'gazed into the mirror and saw what they wanted to see.

They saw their right hand scratching their right ear and we, humanity, followed their lead.

We, humanity, assumed what they had to say was true and we followed in their footsteps as if the truths they espoused were in fact 'truths' when in fact they were simply perceptions, their own personal perceptions, as best they were able to describe them.

We assumed what they had to say was fact when in fact it was simply the best they could express the facts they had available to them personally.

We know in part, and we prophesy in part.

But when that which is perfect is come, then that which is in part shall be done away.

When I was a child, I spake as a child, I understood as a child, I thought as a child: but when I became a man, I put away childish things.

For now we see through a glass, darkly, but then face to face: now I know in part: but then shall I know even as also I am known[ix].

To look in a mirror is not to look in a mirror but to look at the inverse of the image, the left becomes the right and the right becomes the left.

Looking into nothingness does not cause us to see into what lies beyond the barrier of our reality but rather allows us to see a vision of what lies there.

To introduce a new concept into the picture, it will be helpful to revert to a simpler diagram than the ones that have been evolving.

Panentheism
Addressing
Anthropocentrism

As such, to understand the concept regarding the 'location' of nothingness we will revert to the diagram:

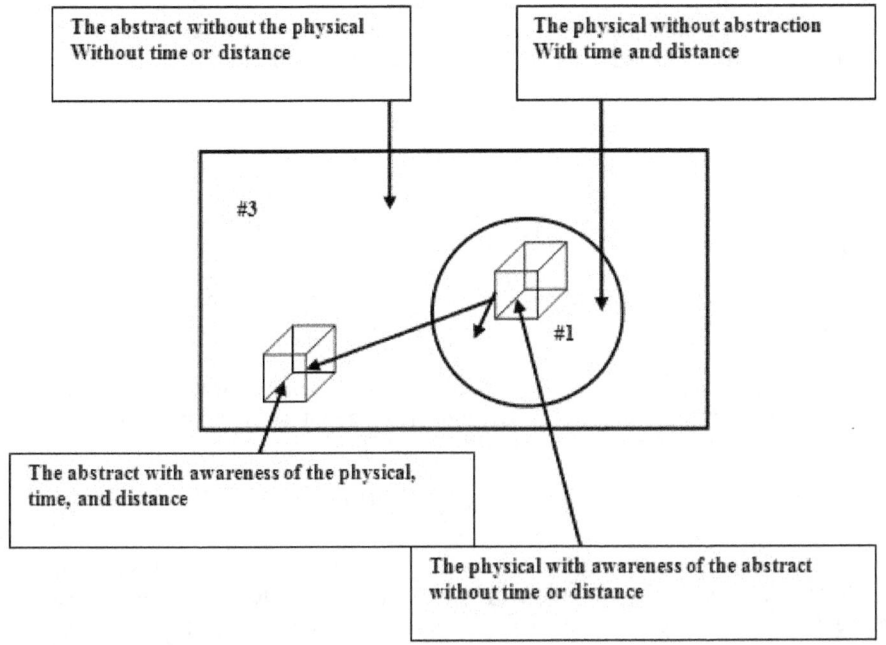

The question now becomes: If nothingness exists, then where 'could' it exist? To understand such a discussion it becomes necessary to answer three questions:

1. Does 'nothingness' exist?
2. What is 'nothingness'?
3. Where is 'nothingness' found?

Why this order to the questions?

The concept regarding the very existence of nothingness evolves even before we define it because it was Boethius who suggested:

'The cause of this mistake is that people think that the totality of their knowledge depends on the nature and capacity to be known of the objects of knowledge.

But this is all wrong.'...

'The point of greatest importance here is this: the superior manner of knowledge includes the inferior, but it is quite impossible for the inferior to rise to the superior.'...

'In the same way, human reason refuses to believe that divine intelligence can see the future in any other way except that in which human reason has knowledge.'…

'… it is quite impossible for the inferior to rise to the superior.', which in turn implies it is quite illogical for the 'inferior', 'ourselves', to perceive of either a concept or 'something' which is imperceptible to the 'superior'[x],

1. Does 'nothingness' exist?

We perceive of 'nothingness' as the lack of 'something'.

Since we perceive of 'nothingness' as a lack of even abstraction itself, it would seem it exists.

The very removal of 'nothingness' undermines all our present day fundamental cosmological, ontological, and metaphysical debates starting with Boethius himself.

Is the existence of 'nothingness' itself an absolute?

Strangely enough, removing 'nothingness' from our present metaphysical model leads to the eventual termination of the system itself, leaving 'nothingness' in place of the metaphysical model we removed. In such a scenario, 'nothingness' becomes 'something'.

In essence, a paradox arises equaling that which Boethius wrestled, equaling the paradox regarding free will versus divine foreknowledge.

Within the perspective of the new metaphysical system of the individual 'acting within'/being a part of God, removing 'nothingness' leaves the system intact.

Panentheism
Addressing
Anthropocentrism

Thus, under such a scenario, 'nothing' becomes just that 'nothing'. Within this type of metaphysical system, NO paradox arises.

2. What is 'nothingness'?

Nothingness is nothing. Nothingness is a void of all.

1. Nothing is a lack of matter.

Since matter is perceived to be 'something'. Nothingness, therefore, must 'contain' no matter.

2. Nothing is a lack of energy.

With the development of Einstein's equation demonstrating a direct relationship between matter and energy, energy is perceived to be 'something'.

Nothingness, therefore, must 'contain' no energy.

3. Abstraction is perceived 'to be'.

Nothingness is the lack of all, a void of all. If abstractions are a 'part' of nothingness, then nothingness is no longer 'nothingness'.

Any existence capable of being subdivided into sub-parts is equal to the sum of its parts.

As such, any existence capable of being subdivided is by definition not 'nothingness' for by definition it is no longer a 'void' of all but rather a sum of all.

What then is 'nothingness'? Nothingness is definitely not matter.

Nothingness is not energy. Nothingness does not appear to be abstraction. It appears nothingness is closer to being energy than matter.

Nothingness appears to be even closer to what we perceive to be abstraction than it is to being energy.

In all likelihood however, nothingness may be our first glimpse into what acts as the fundamental unit of existence itself.

It is easy to come off task at this point. Moving into a dialectic regarding the topic of nothingness being the fundamental unit of existence itself, becomes an almost irresistible act.

However, our task - within this tractate - is to explore the concept of Centrism versus non-Centrism.

As such, we must return to work.

3. Where is 'nothingness' found?

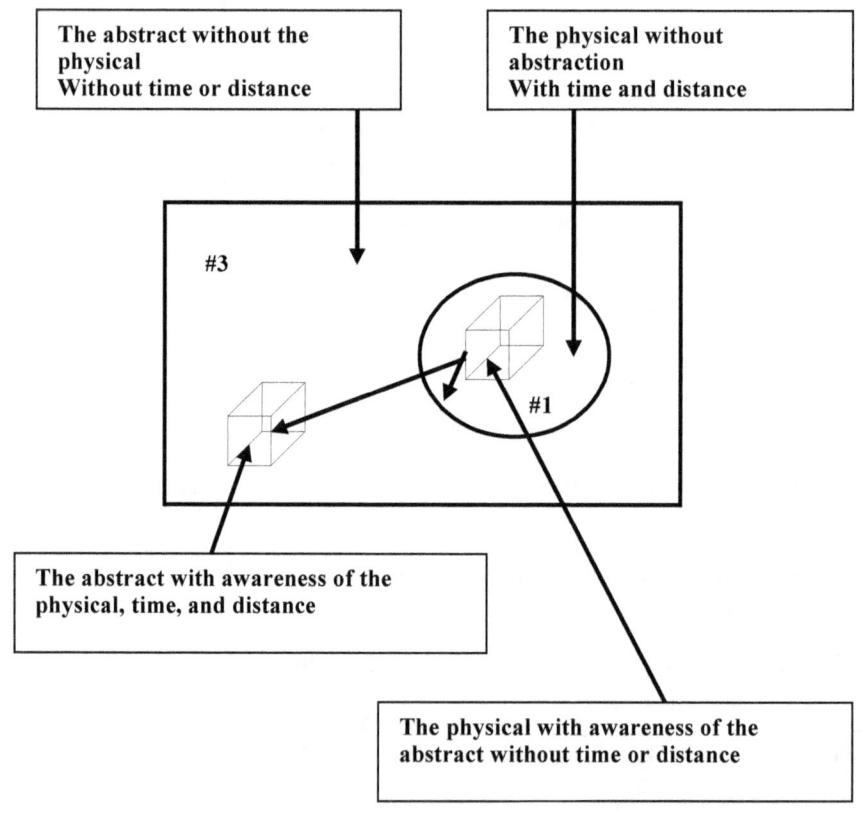

Panentheism
Addressing
Anthropocentrism

Questions arise:

1. Can 'nothingness' be found 'outside' region #3?
2. Can 'nothingness' be found 'outside' region #1?
3. Can 'nothingness' be found 'within' region #3?
4. Can 'nothingness' be found 'within' region #1?

In terms of question #1:

Can 'nothingness' be found 'outside' the whole, 'outside' region #3?

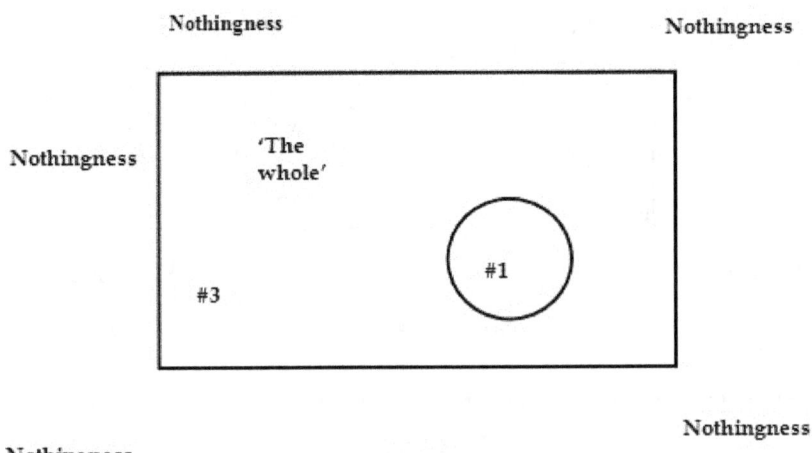

Daniel J Shepard

Channel

Since 'nothingness' in this graphic 'lies' 'outside' 'the whole' the whole to be the whole must expand to include what lies beyond it to remain 'the whole'.

As such we obtain:

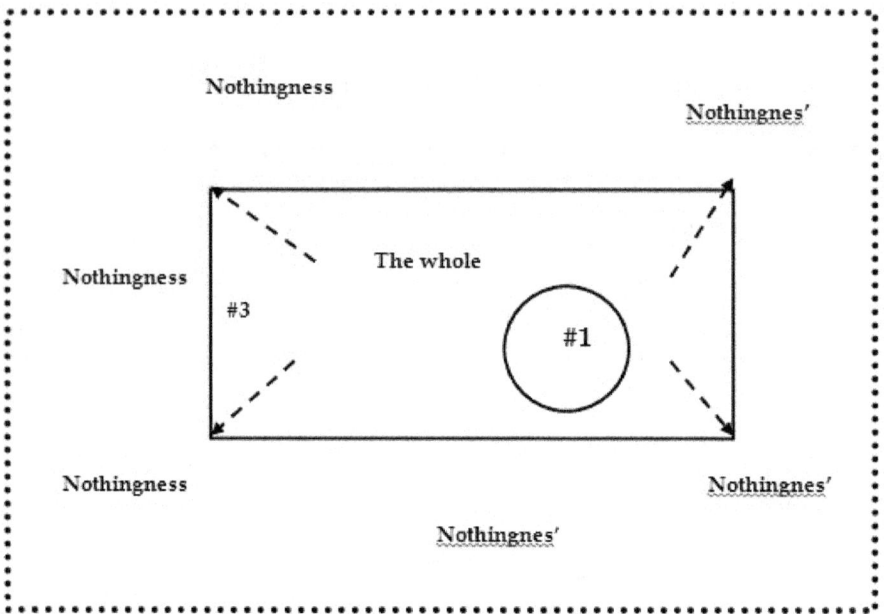

Panentheism
Addressing
Anthropocentrism

As such the 'whole' now becomes:

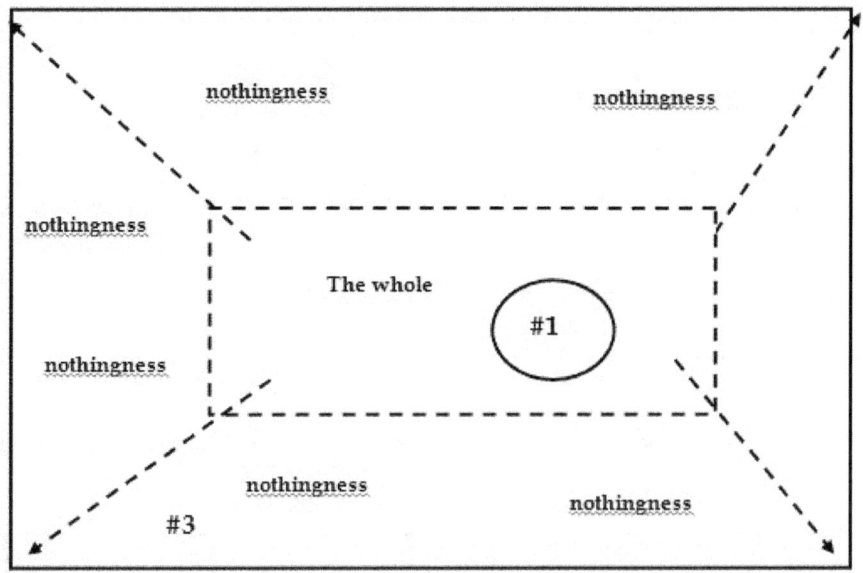

The significant result is that it appears 'a whole without 'nothingness'' is different from 'a whole with 'nothingness''.

If 'nothingness' exists, it appears to 'lie' 'within' 'the whole'.

If one places 'nothingness' 'outside' the 'whole', the 'whole' must expand to include 'nothingness' or it is, by definition, no longer the 'whole'

Therefore in terms of the question: Can 'nothingness' be found 'outside' the whole, 'outside' region #3? The answer is 'nothingness' does not exist 'outside' the whole at least from all possible perspectives of the whole itself.

If such is the case, the question arises: Does 'nothingness' have a function?

As we are about to explore, nothing lacks function within the whole, including 'nothingness' itself.

In terms of question #4:

Can 'nothingness' be found 'inside' the physical, 'within' region #1?

Is this not out of order? Alphabetically, yes it is out of order.

Order, however, is not the point here. In fact what we are about to do is examine the concept beginning with the 'furthest' 'outward' reaches of which we can possibly conceive, followed by the furthest 'inward' reaches we can possibly conceive.

In short, we are examining the furthest extremes conceivable for 'nothingness'.

Having done so we will then examine 'nothingness' in terms of the more moderate points of view.

We will begin exploring the inner most extremes of 'nothingness with our last graphic.

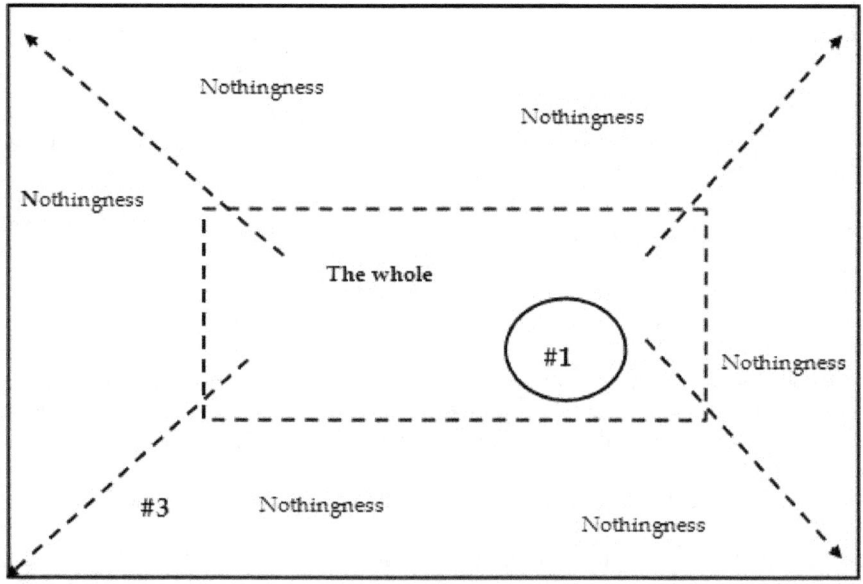

Panentheism
Addressing
Anthropocentrism

To better depict the process of examining 'nothingness' in terms of the inner most extreme, we will remove the 'outer' aspect of our drawing and enlarge region #1:

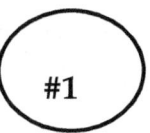

We will not enlarge region #1:

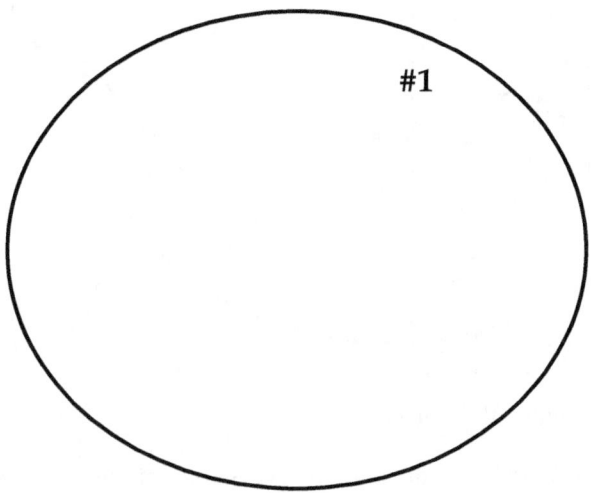

Daniel J Shepard
Channel

We recognize this as the physical.

We often label this region, the universe.

Taking the region for what it is we will label it and place a representative of ...

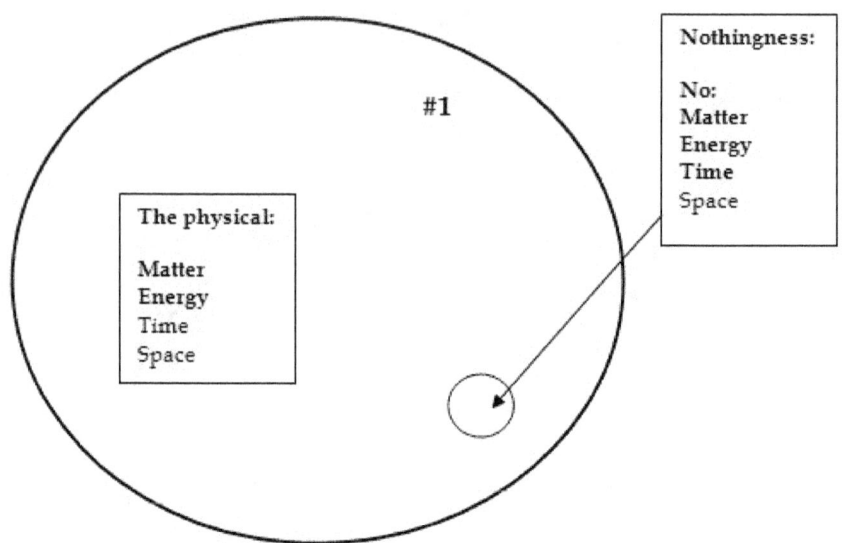

... nothingness within this universe.

A vacuum is unnatural state of nature but we are not discussing a vacuum, we are discussing a region of nothingness.

Not only is nothingness a lack of matter, but also nothingness is a lack of energy.

We can move to an even greater extreme, for nothingness is not only the lack of matter and energy but nothingness is a lack of matter, energy, and abstractions.

Thus time and space would be absent within a region of nothingness.

Some would argue the lack of space and time cannot be found anywhere within the region of the universe and therefore such a concept is irrelevant.

Time and space appear to be an innate characteristic of matter and energy.

Some would disagree and say matter and energy are innate characteristics of time and energy.

Panentheism
Addressing
Anthropocentrism

This is the old puzzle: which came first – the chicken or the egg?

In this particular puzzle, the subject deals with innate characteristic and as such becomes:

Are time and space innate characteristics of matter and energy, or matter and energy innate characteristics of time and space?

Since time and space are missing 'within' the region of 'nothingness', the region takes up no time and space and thus is in essence non-existent from the perspective of the region of the physical.

Therefore in terms of the question: Can 'nothingness' be found 'inside' the physical, 'within' region #1?

The answer is 'nothingness' does not exist 'inside' the physical at least from all possible perspectives of the physical itself.

It might be feasibly possible to either find or create a region in the universe lacking matter and energy but it does not appear to be feasibly possible to find or create a region in the universe lacking time and space for time and space appear to be the very fabric of the universe itself.

Daniel J Shepard
Channel

In terms of question #3:

Can 'nothingness' be found 'inside' the whole, 'within' region #3?

To better understand the answer to this question we must refer back to a diagram found in section: The dynamics of non-Centrism:

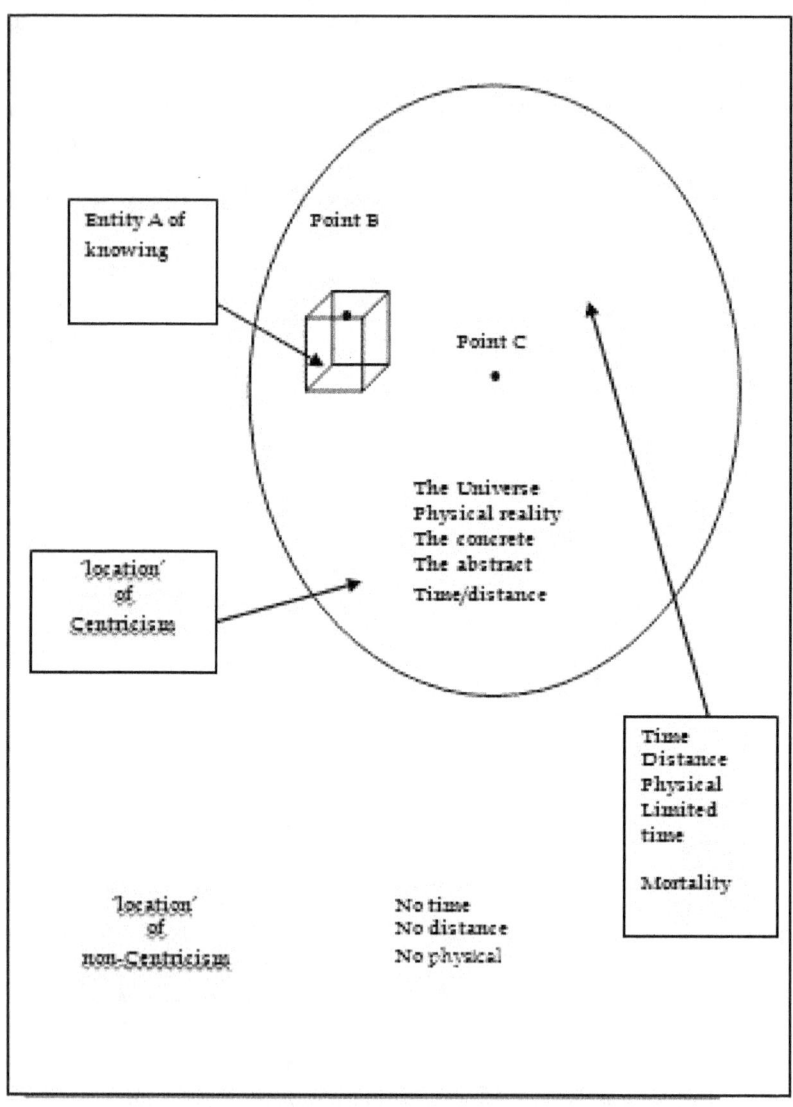

Panentheism
Addressing
Anthropocentrism

We will 'reduce' the graphic to a more manageable form and then add 'nothingness' within region 3:

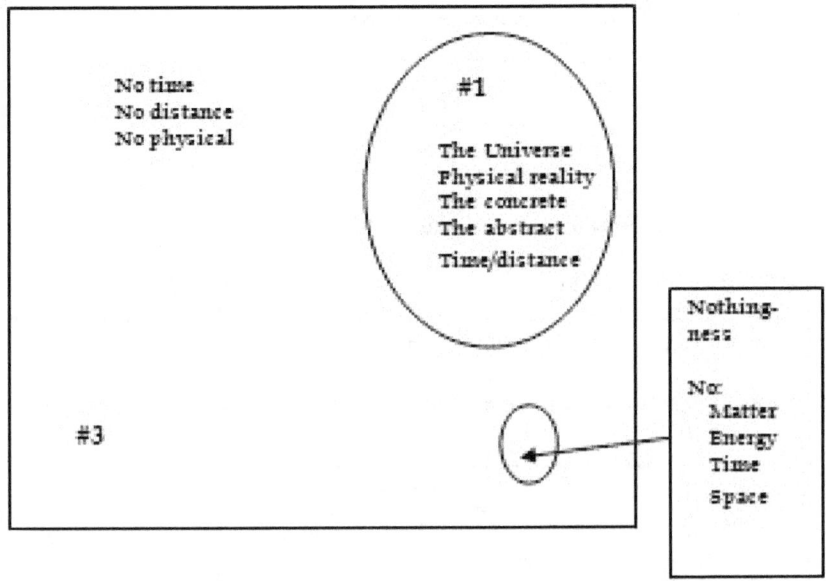

This concept is understandable if one acknowledges an 'existence' of the physical lying within the non-physical, lying within the abstract.

The question now becomes: Does abstraction lie within 'nothingness'?

The answer would appear to be: If 'nothingness' is truly 'nothingness' then abstraction itself would be found to be absent 'within' 'nothingness'.

Could such a region exist 'outside' the physical but 'within' the abstract?

The answer appears to be: Yes it is conceivable for the lack of matter, energy, time, space, and all abstraction to exist within a region of abstraction itself.

Therefore in terms of the question: Can 'nothingness' be found 'inside' the whole, 'within' region #3?

The answer is 'nothingness' does not exist 'inside' the physical at least from all possible perspectives of the physical itself but 'nothingness' could lie within abstraction, could lie within region #3 if it lies outside region #1.

In short, it might be feasibly possible to either find or create a region within pure abstraction lacking matter, energy, time, and space since time and space are not what the fabric of which the abstract is composed

In terms of question #2:

Can 'nothingness' be found 'outside' the physical, 'outside' region #1?

At first glance, this would appear to be the same question as 'b': Can 'nothingness' be found 'inside' the whole, 'within' region #3?' Questions 'c' and 'b' however, are significantly different questions.

Again, graphs will simplify our understanding of the question.

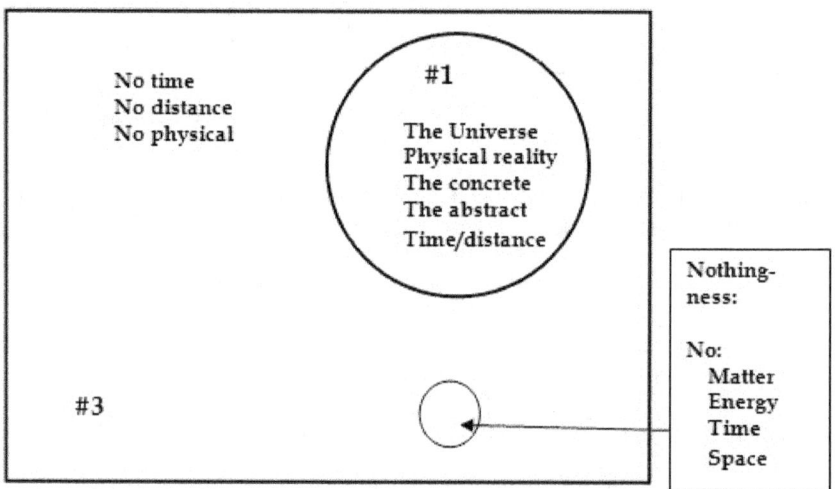

Before we go any further, let's reiterate a concept presented in Volume 5: Zeno.

The concept: A circle divides existence into three regions: The inside of the circle, the outside of the circle, and the circle itself.

Panentheism
Addressing
Anthropocentrism

We must also reiterate two basic principles of geometry: The first concept: A circle is composed of an infinite number of points in a plane located equidistant from 'a' point, the center.

The second concept: 'a' point is a location in space having no dimensions, having no length, breadth, or depth

With this in mind, we will compress the contents of the circle, compress the universe, and apply the contraction aspect of the Big Bang Theory to our diagram.

If we compress the border, separating the physical and the abstract we begin to see the expansion of a region lacking matter, energy, time, and space.

In fact, if we follow the logic of 'b'. Can 'nothingness' be found 'outside' the physical, 'outside' region #1?

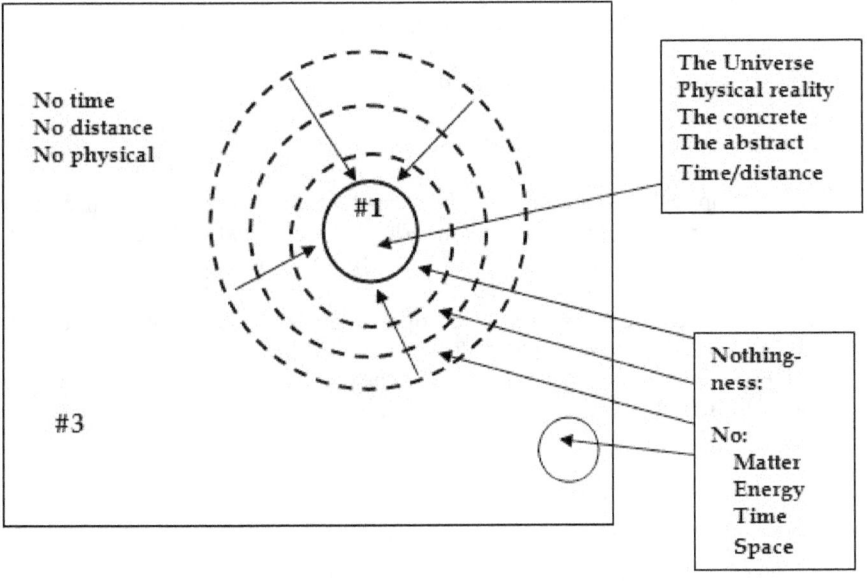

Now we can understand that no matter how much the universe expands or contracts, it remains immersed within a bed of nothingness and therefore takes up 'no more space' if it becomes infinitely large than it does if it becomes infinitely small. The reason for this is that it lies 'within' 'nothingness' which 'contains' no time or space, which lacks time and lacks space.

But wouldn't the universe then lie directly in the abstract?

No, for time and space are abstractual and as we have previously demonstrated in this Tractate, 'nothingness' is not only the lack of matter and energy but 'nothingness' is the lack of all abstractions.

The significance of such a concept now emerges:

The region 'separating' the physical from the abstract has no dimension. Matter, energy, time and space do not exist 'within' 'nothingness' and therefore the boundary separating the physical and the abstract is, from the perspective of the physical as well as from the perspective of the abstract, non-existent.

Now we see that not only can 'nothingness' be found within the whole but also in terms of the universe, in terms of reality, in terms of the physical that contains time and space, 'nothingness' can be found in two places:

1. Adjacent to and therefore separating the universe, separating reality, separating the physical, separating multiplicity, separating Cartesianism, separating free will, separating Centrism, separating all aspects of the physical from the abstract

2. Non-Adjacent to the universe

Both 1 and 2 have their own implications.

Statement 1 implies 'nothingness' is the mirror separating the physical and the abstract. Statement 1 suggests a 'location' where the physical and the abstract face one another just as one faces oneself when gazing into a mirror.

Statement 2 implies the existence of multiple 'realities'. Statement 2 implies the potential of universes existing which do not have time and space as the fabrics of their universes but rather the fabrics of other universes may or may not be composed of other abstractual concepts than the abstractual concepts of space and time.

Panentheism
Addressing
Anthropocentrism

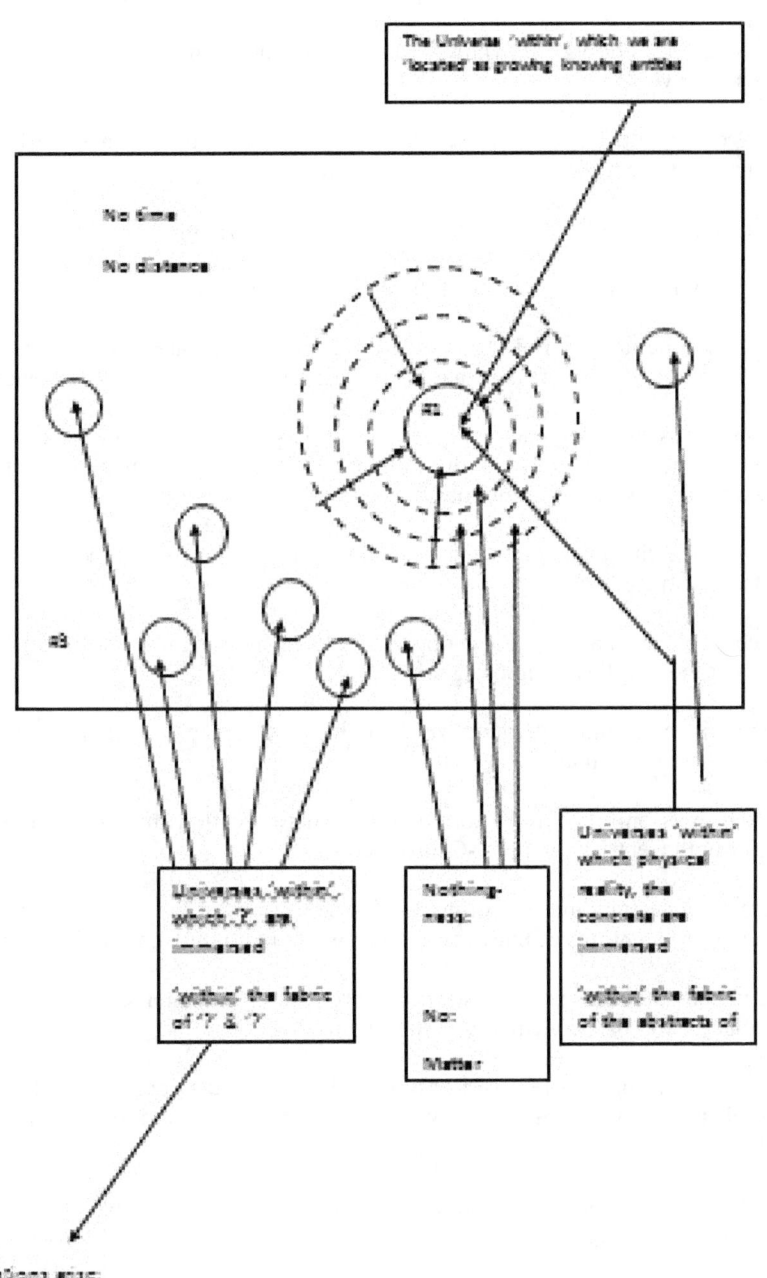

Daniel J Shepard
Channel

Is the first question mark replaced by the physical? It may be but it need not be.

This capability allows the whole to be much more complex than if the answer were simply yes.

The second question, which arises, is: Are the remaining two question marks the abstractual concepts of time and distance/space?

Again the answer is the same, they may be or they may not be.

Is there anything we can say about the three question marks?

Yes, we can be fairly certain the last two question marks are forms of abstractions since the reality of universes is that they all lie 'within' abstractual existence.

The questions then become: Is either the basic concept of Centrism or the basic concept of non-Centrism or both simultaneously for that matter, found to be an innate characteristic of 'nothingness' itself?

Centrism is a characteristic. As such Centrism must apply to 'some' concept be it physical or abstractual.

The concept of 'characteristics' do not apply to 'nothingness' rather the void of characteristics applies to 'nothingness'.

Characteristics cannot 'characterize' 'nothingness', since 'nothingness' is neither physical nor abstractual.

As such, it is only 'within' 'nothingness' where we find the lack of one or the other or both Centrism and non-Centrism.

Where do understandings regarding the characteristics of 'inverse proportionality', 'nothingness', 'entities of knowing', 'Centrism', and 'non-Centrism' lead us?

Such understandings lead us toward comprehending the interaction as well as the interrelationship of the elements of the 'whole'.

Since by definition we are a part of the whole, such understandings lead us toward comprehending the interaction as well as the interrelationship of ourselves to the whole.

Before we can explore such interrelationships and interactions, we have two other concepts, which need addressing.

Panentheism
Addressing
Anthropocentrism

13. Virgin physicality/'virgin physical life'

The concept: 'Nothingness' found 'within' the physical, found 'within' region #1 is 'Centrism' itself.

In fact, 'nothingness' found 'within' the physical is the 'fundamental building block' of the physical, is a concept for the field of science/cosmology to explore.

We find ourselves in the middle of a dialectic regarding the differences of existing within a region of Centrism (region #1) and a region of non-Centrism (region #3).

The similarities of the two regions, as well as the significance one region imposes upon the other, are the concepts we are attempting to understand.

We can begin our understanding by making a simple statement:

'Nothingness' found 'within' the physical, found 'within' region #1, is Centrism itself.

To understand the concept: 'Nothingness' found 'within' the physical, found 'within' region #1, is Centrism itself, we will once again rely upon graphics:

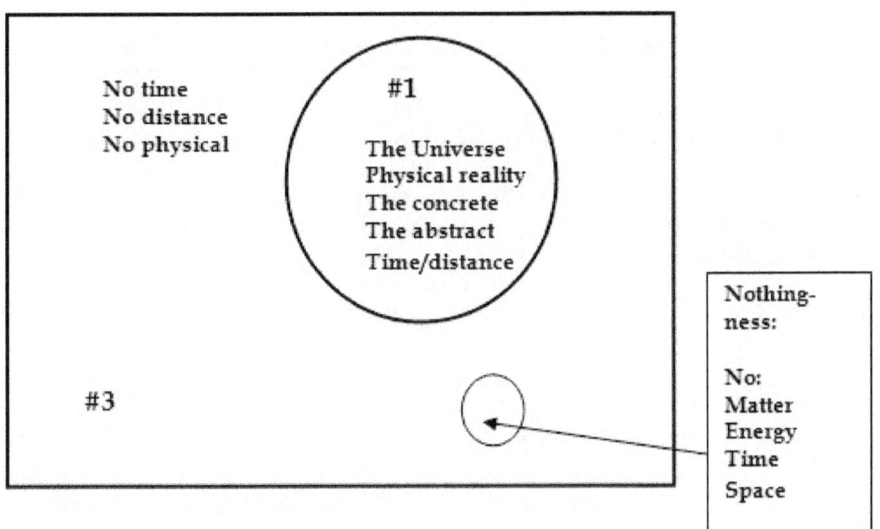

We will now expand the universe to include 'nothingness':

Daniel J Shepard

Channel

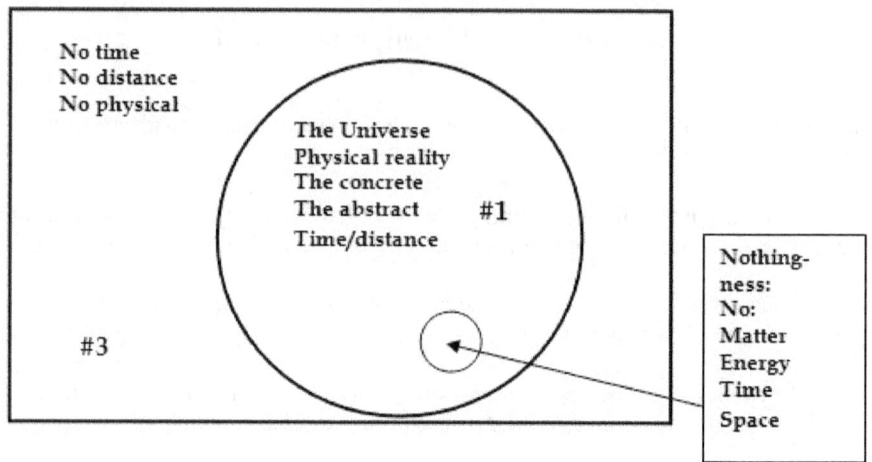

As we have discussed previously, nature abhors a vacuum and as such rushes to fill the vacuum.

But nature acts symmetrically and as such nature rushes to fill the vacuum equally from all 'directions' until the vacuum is no longer a vacuum.

The point is the vacuum collapses from the 'outer' edge toward the 'center'.

Within the physical, 'nothingness' is the location towards which all 'things' move to find the center.

'Nothingness' becomes the epitome of the center itself.

'Nothingness' becomes the center for as 'nothingness' becomes 'occupied' and its radius diminishes to non-existence itself, 'nothing' has occurred no 'space' was lost and no 'distance' traversed.

Since no distance was traversed, no time was taken to fill the region of 'nothingness'.

This statement can be made when referring to time for time is not an element, is not a part of the fabric of which 'nothingness' is composed since by definition 'nothingness' is the void of all including not only matter and energy but also distance and time.

What then is: 'Virgin physicality/'virgin physical life'?

'Virgin physical life' is physical life at the point of centralism's formation. 'Virgin physicality' begins just before formation but just after it pre-existed as separation.

Panentheism
Addressing
Anthropocentrism

'Virgin physical life' is the point of origination, is physical life 'before' it has gained the ability to physically function as singularity of existence versus multiple independent existence of components.

'Virgin physicality' is physical existence as 'a' unit versus existence as separate components 'capable' of combining to form 'a' unit.

'Virgin physical existence' is in essence nonexistence, is in essence nothing, is in essence 'nothingness' itself.

'Virgin physicality/'virgin physical life' is the moment 'before' space, time, matter, energy forms of experience become a part of knowing and thus knows nothing of its own self for its own self has yet to begin its process of knowing.

The lack of knowing existing immediately before the process of knowing begins is where 'virgin-ness' lies.

Virgin-ness lies at the center of 'nothingness' itself. Cosmologically and ontologically,

Centrism thus finds its ultimate source of centricity.

What then of the universe?

Science and religion with their principles of symmetry and creation would suggest the same concept can be applied to this thing we call 'virgin physicality'.

Again, however, this is a discourse for science/cosmology and religion/ontology to explore.

Our task is metaphysical in nature and although our discussion may have direct implication for science and religion, it is not the scientific nor the religious aspects we are to pursue.

What then are we attempting to explore?

We are attempting to explore the concept of knowledge becoming 'knowing' knowledge, consciousness of 'knowledge'.

Channel

We will, therefore, modify our graphic by replacing the symmetrical circle within which we find three dimensions, four dimensions, and may in the future find five, six, and more dimensions with a unit entity of 'knowing': In short, we are about to explore the concept of: Virgin consciousness/'virgin abstract knowing'.

Panentheism
Addressing
Anthropocentrism

14. Virgin consciousness/'virgin abstract knowing'

We can now make a second statement regarding the concept of 'nothingness'

'Nothingness' found 'within' the abstract, found 'within' region #3 is 'non-Centrism' itself.

It could be said that the statement: 'Nothingness' found 'within' the abstract is the 'fundamental building block' of the abstract.' is a concept for the field of religion/ontology to explore.

We find ourselves in the middle of a dialect regarding the differences of existing within a region of Centrism (region #1) as explained by cosmology and existing in a region of non-Centrism (region #3) as explained by ontology.

This is the very point of metaphysics. Metaphysics listens to both cosmology and ontology and attempts to resolve the issues keeping the two apart.

Before we can make the leap of resolving the basic differences between ontology and cosmology, we must allow the metaphysical understanding regarding the ontological perception of 'nothingness' to emerge just as we allowed the metaphysical understanding regarding the cosmological perception of 'nothingness' to emerge.

To understand the concept: 'Nothingness' found 'within' the abstract, found 'within' region #3, is non-Centrism itself.' we will once again rely upon graphics:

Daniel J Shepard
Channel

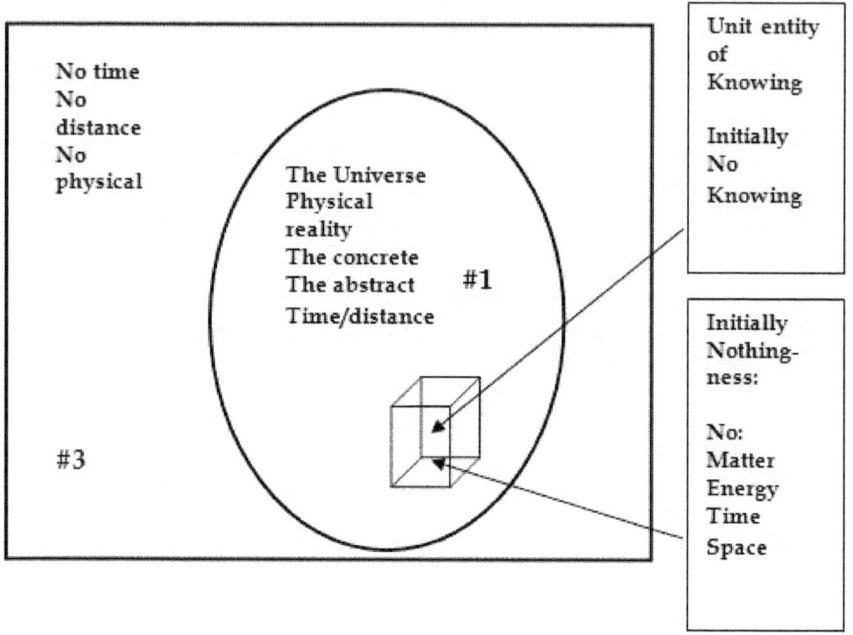

As the entity moves through time and space, it gains an 'awareness' of experiencing.

The entity begins to 'fill' up with both knowing and awareness of its knowing. Eventually, the entity becomes 'full'.

The 'fullness' of the entity does not terminate the entity's existence in the physical.

Rather the termination of the entity's existence in the physical completes the filling of the entity.

This does not imply the entity has 'room' for more knowing, for the entity is always 'full'.

Rather the physical termination is the termination of the 'process' of acquiring knowledge by the knowing of the entity.

Panentheism
Addressing
Anthropocentrism

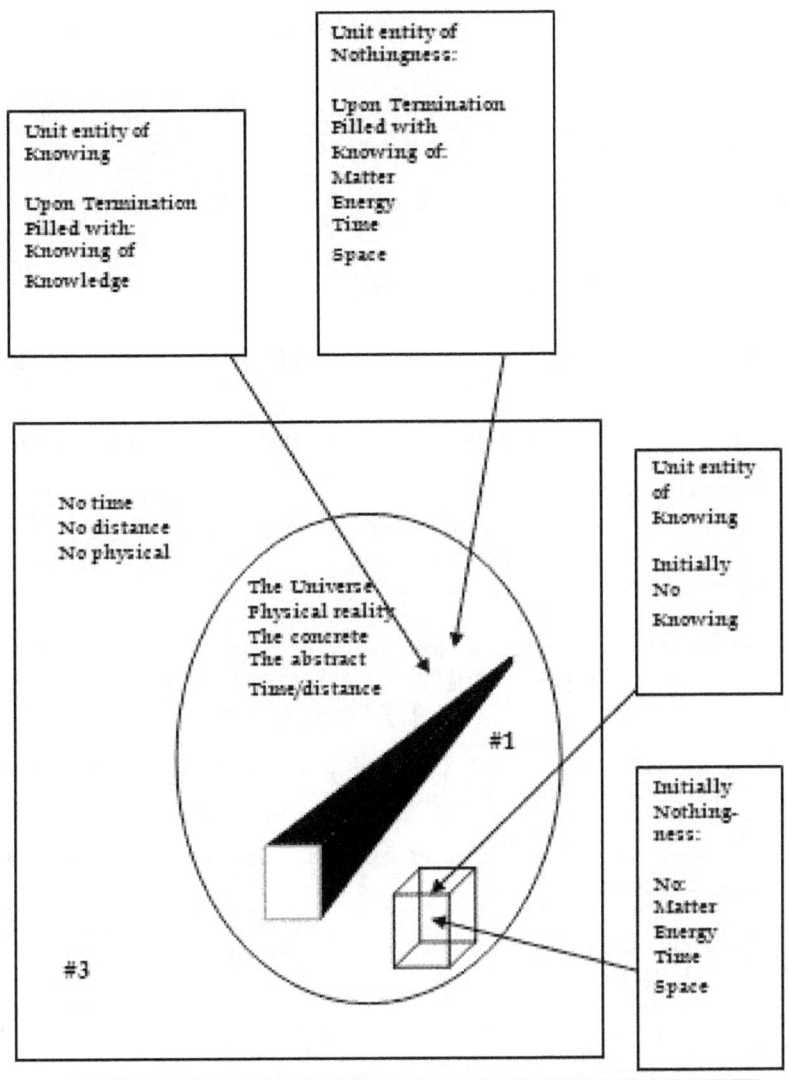

We can depict the completion of the process as a 'filled' versus an 'empty' unit.

Regardless of the 'amount' of knowledge 'contained' within the entity of knowing, the concept of the entity being completely filled is complete with the termination of gaining knowing unique to the entity of knowing.

Likewise, regardless of the 'amount' of knowledge 'contained' within the entity of knowing the concept of the entity being completely empty while in the process of gaining knowing unique to the entity of knowing is in essence 'empty' relative to the potential to gain knowledge by the entity of knowing.

The only limit imposed to the continued increase in knowledge forced upon the unit of knowing is the termination of the process itself.

Some would call this termination of the growth process of 'knowledge' by the unit of knowing, death.

There is nothing inappropriate regarding this label.

However, to say we 'know' what death is, is inappropriate for we can only speculate regarding such a meaning.

In this Tractate, we are exploring the very meaning of the term death, from a different direction than that which society presently comes.

We are exploring the perception of death - exploring the perception of the mirror separating the physical and the abstract from the point of view of a new metaphysical perception.

What then is: Virgin consciousness/'virgin abstract knowing', 'virgin abstract knowing' is knowing at its point of origination, is knowing 'before' it has gained any knowledge and thus its knowing knows nothing.

Virgin consciousness/'virgin abstract knowing' is the moment 'before' space, time, matter, and energy forms of experience become a part of knowing.

Virgin consciousness knows nothing of its own self for its own self has yet to begin its process of knowing.

The lack of knowing existing immediately before the process of knowing begins is where 'virgin-ness' lies.

Virgin-ness lies at the center of the 'nothingness' of knowing itself. Metaphysically, Centrism thus finds its ultimate source of centricity.

Panentheism
Addressing
Anthropocentrism

15. Stepping 'in' beyond Centrism: Dependency

Now as much as it may appear logical to begin examining the entities of knowing within region #1, we are going to begin examining the entities of knowing from region #3.

To do this we will simply move the entities of knowing out of region #1 and into region #3.

In essence, we are going to step 'out' into non-Centrism. We are going to allow the two forms of unique knowing of knowledge to step 'out' into non-Centrism.

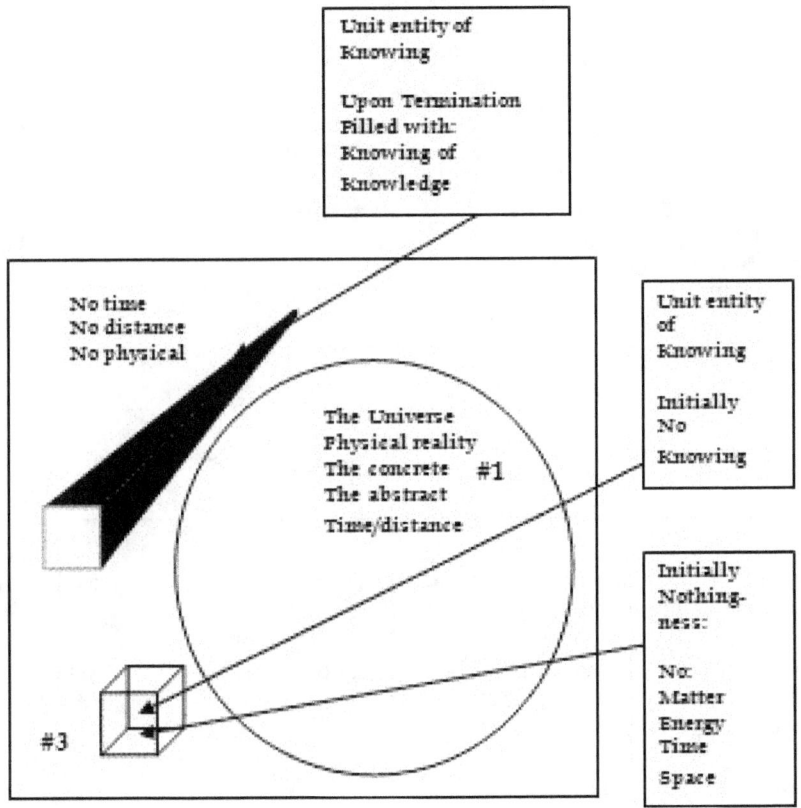

Since the 'empty' unit, the 'virgin entity of knowing', has absolutely nothing abstractual or otherwise within it, it takes up no space or time and thus we can now simplify the graphic:

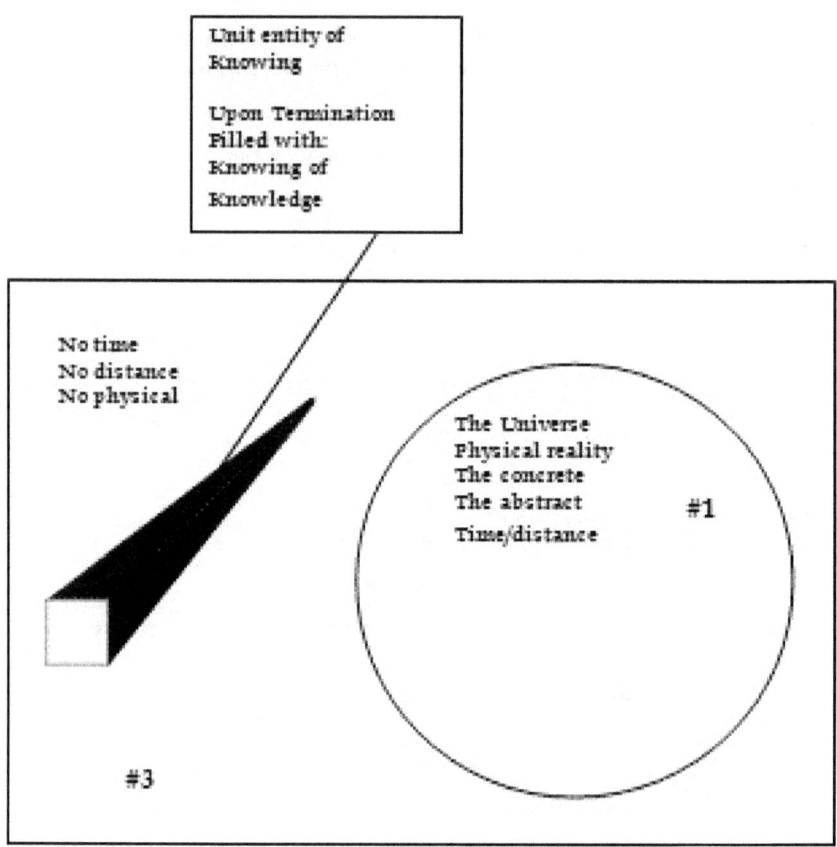

With the aid of this graphic, we can now begin to understand the interrelationship between existing entities found within region #3.

We can now begin to understand the very concept of not only the very existence of non-Centrism itself but we can begin to understand the process as well as the potentiality of non-Centrism.

Existence within non-Centrism:

Panentheism
Addressing
Anthropocentrism

We will take the latter diagram and reduce region #1, the location of Centrism since we are concerned with region #3.

In addition we will expand upon the number of entities of knowing which have evolved through the process of 'traveling', experiencing region #1 – the location of space, time, matter, energy…

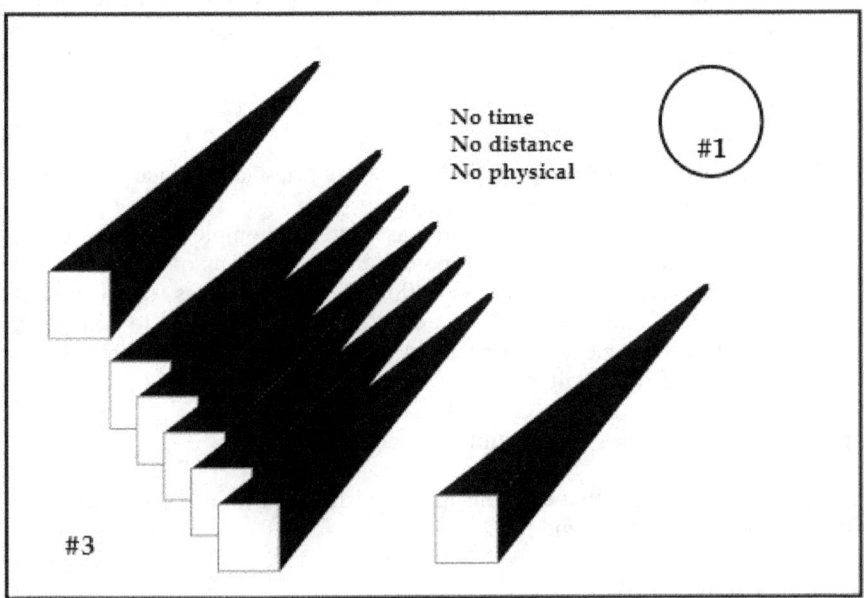

The entities of knowing:

1. Are each unique in and of themselves due to their unique experiencing and assembly through time and space or whichever abstractual fabric should exist within the 'universe' from which they emerge
2. Do not vary in size since relative to the whole, each is relatively the same in size
3. May appear to vary regarding their 'distance' apart but since they are immersed in a location void distance, in essence the 'distance' separating them does not exist. They are, therefore, all equidistant – no distance apart
4. Are independent of beginning – end parameters since they are immersed in a void of time and space, immersed within a form of non-Centrism

Process

> 5. Take no 'time' to get from one to another since there is no 'distance' to traverse
> 6. Are capable of 'knowing' one another completely since they are not knowledge but passive unique experiencing of knowledge assembled by the active process of knowing and whose process remains an integral part of themselves since the passivity of knowledge is nothing without active process of knowing

Potentiality

> 7. Are capable of incorporating 'new' entities of unique knowledge, knowing, and experiencing
> 8. Partial summations of knowing as well as complete summations of knowing are as varied as the potential combinations of existing entities and the parts of existing entities allow. This is known as 'Omniscience'.
> 9. Potentiality within region #3 is, due to #1 – 8 is limited to what is and therefore is 'dependent' one entity upon another entity, which exists as opposed to what 'might' exist.

Dependence thus becomes the principle of region #3.

It must be acknowledged that the concept of existence 'within' non-Centrism is a difficult concept to comprehend.

But why shouldn't it be?

We are after all immersed 'within' Centrism' and as such find time and space to be concepts expanding 'outward' from 'a' point of reference which varies from conscious knowing to conscious knowing.

What must not be lost within the exploration of the unfamiliar, however, is that within a location void the fabric of space and time, 'correct' sequencing is not a fundamental principle since 'correct' sequencing is an aspect of time and space found within our personal universe.

Region #3 therefore is a location where our unique experiencing depends upon what others have to offer us to experience versus our forming our own experiencing for our own unique experiences are what they are and our own uniqueness cannot change without the infusion, embracing, union of another's unique knowing.

Divine intervention?

Panentheism
Addressing
Anthropocentrism

Perhaps.

Who is to say what the whole is capable of doing with its creation: the universe, our reality.

Nevertheless, how can an 'all knowing' entity be all knowing if it doesn't know 'all things'?

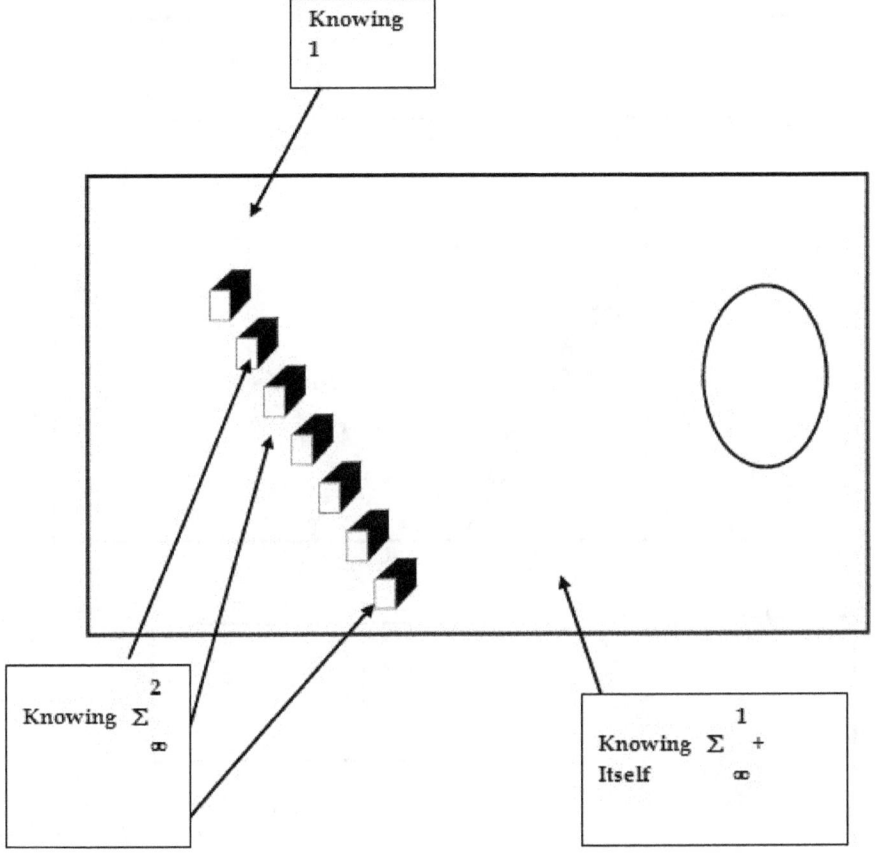

Doesn't this diagram imply 'all Knowing' doesn't know what 'will be' and doesn't this in turn imply time exists in 'all Knowing'?

Daniel J Shepard

Channel

The concept of what 'will be' only exists 'within' time and if we review what we had previously learned of time, we find time to be 'located' within two locations:

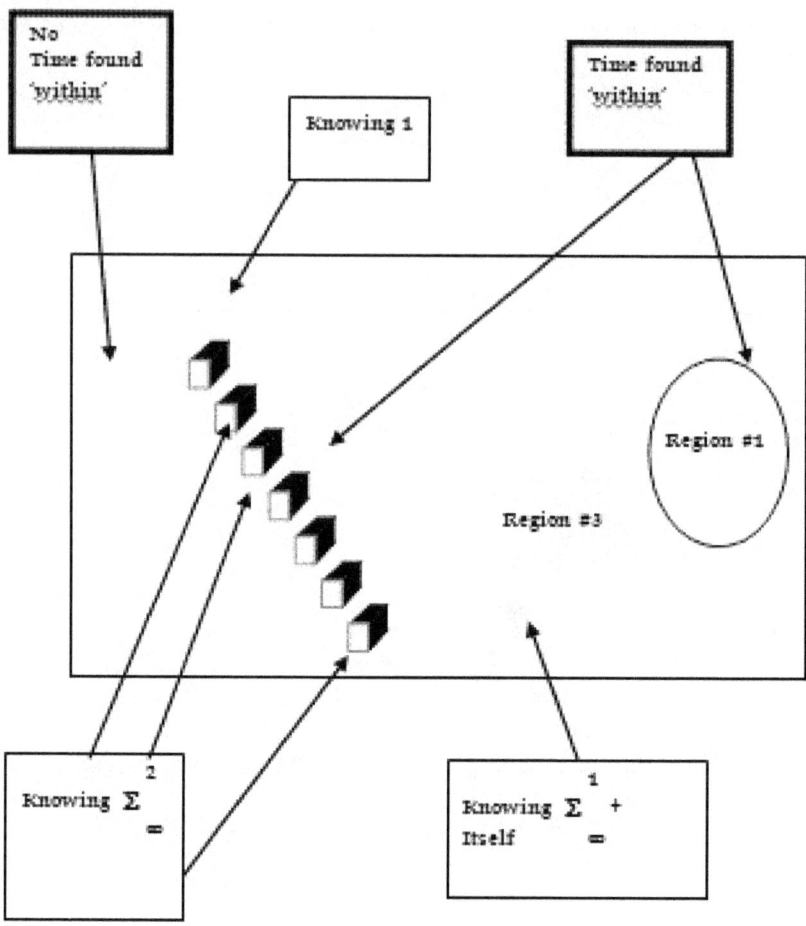

Since time is not found either as an innate characteristic of region #3 nor as a medium of region #3 'within' which subsets of region #3 find themselves immersed, there is no concept of 'what will be' to be found in region #3.

This is not to say time is not found 'within' region #3, rather it demonstrated time is not an all-pervasive characteristic, a universal medium of region #3.

Panentheism
Addressing
Anthropocentrism

Time is found in region #3 in two distinct location of which we are presently capable of perceiving. Time is found 'within' entities of 'knowing', which have evolved out of the 'universe', and time is found as a medium of the universe within which entities of 'knowing' move from being 'virgin' entities of 'knowing' to being complete entities of 'knowing'.

Where then does the 'Book of Divine Knowledge' find itself to be in terms of the metaphysical system of singular location, the individual 'acting within'/being a part of God.

Daniel J Shepard
Channel

Panentheism
Addressing
Anthropocentrism

16. Stepping 'into' Centrism: Independence

Having stepped 'in' beyond Centrism and into the region characterized by non-Centrism, let's now step 'out' of non-Centrism and into Centrism – Independence.

To do so we will begin where we initiated the understanding of region #3:

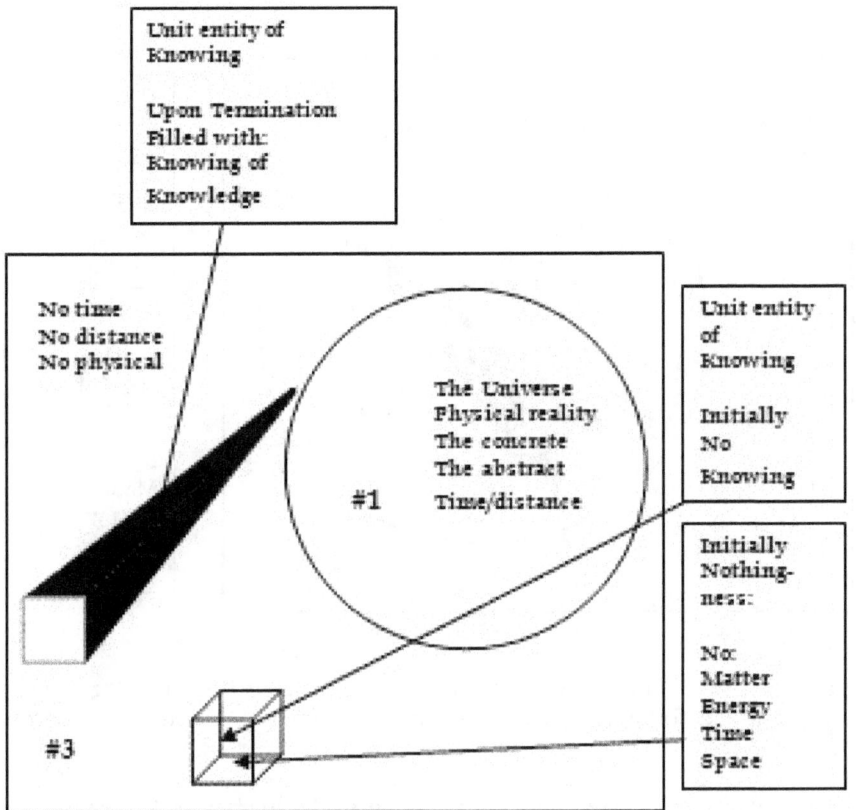

By moving 'into' Centrism we are in essence moving 'beyond' non-Centrism/dependency and into Independency.

As such our graphic now becomes:

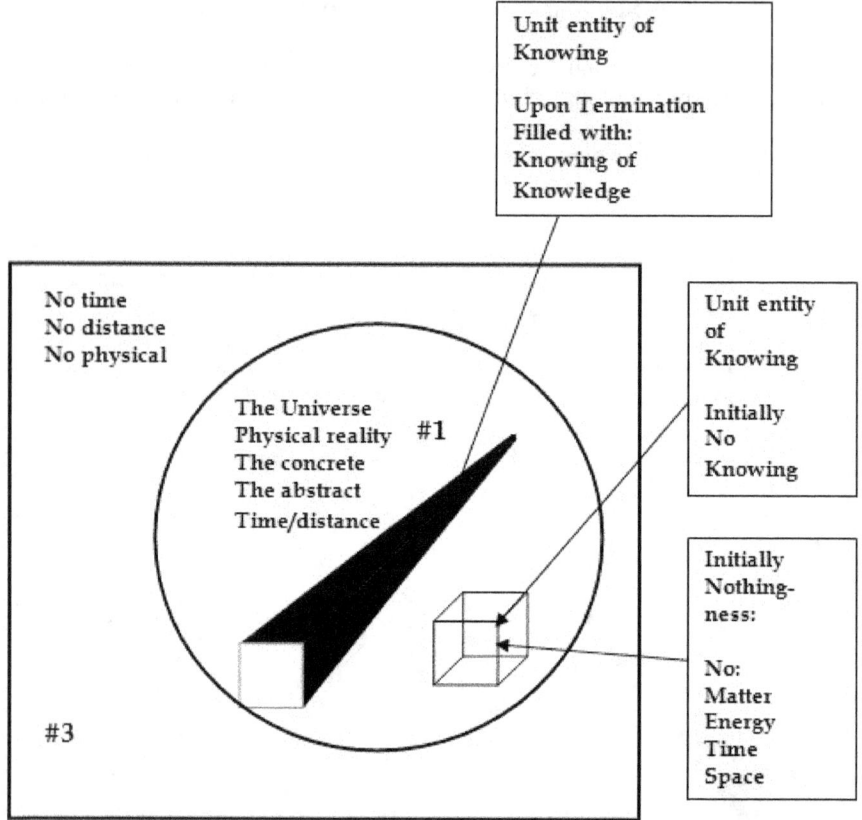

Region #1 is a location where our unique experiencing begins at the point of nothingness and forms in a unique manner to ourselves based upon our own unique experiences.

As we previously discussed, upon termination of the journey of knowing within time and space, the entity is no longer 'empty' or partially 'empty' of knowledge.

This leads us to our next step.

We are going to remove the 'filled' entity of knowing since we are focusing upon the 'process' of knowing and how it interacts with the concept of Centrism, which Copernicus so aptly entrenched in the field of science.

Panentheism
Addressing
Anthropocentrism

As such, our graphic becomes:

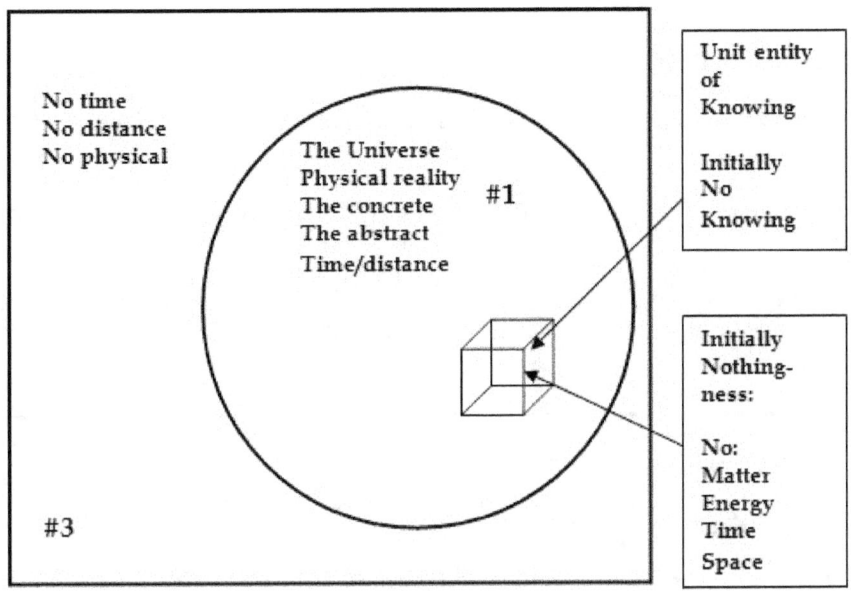

The 'empty' unit, the 'virgin entity of knowing' has absolutely nothing abstractual or otherwise within it at only one point in time.

The 'virgin entity of knowing' is empty when it is simply 'process'.

Until the process of knowing begins, the entity is in a state of passivity and only enters the active state of *being* when it makes the step beyond being in a passive state to being in an active state.

A detailed discussion of such states was explored within Volume 7: Aristotle.

Since we discussed the concept of a passive state of knowing in Volume 7, we will proceed to examine the active state of knowing, the process of filling an abstractual unit of active knowing with the substance it needs to exist as an entity, the substance of knowledge and experience.

The entity exists enmeshed in the fabric of time and space/distance since time and space are the very fabrics of region #1.

With the aid of this graphic, we can now begin to understand the interrelationship between existing entities found within region #1.

Daniel J Shepard
Channel

We can now begin to understand the very concept of not only the very existence of Centrism itself but we can begin to understand the process as well as the potentiality of Centrism.

Existence within Centrism:

We will take the latter diagram and reduce region #3, the location of non-Centrism since we are concerned with region #1.

Since region #1 is located 'within' region #3, the only way to do this while expanding region #1 is to apply the concept of relativistic size of region #1 compared to region #3. In addition to modifying the relative size of each region, we will expand upon the number of entities of knowing which evolve through the process of 'traveling', experiencing region #1 – the location of space, time, matter, energy…

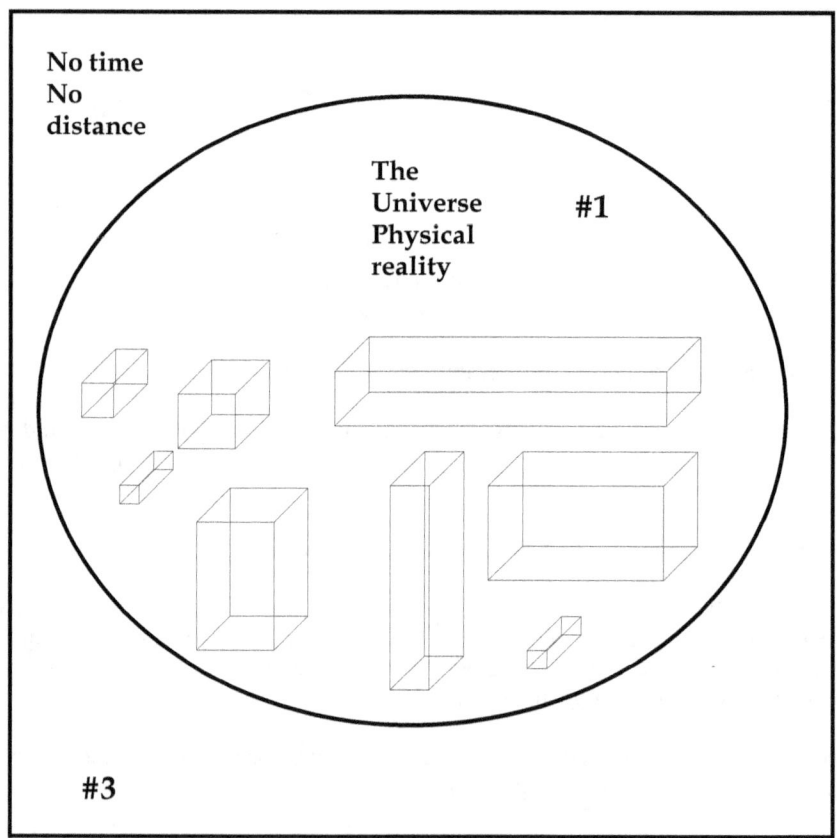

Panentheism
Addressing
Anthropocentrism

The entities of actively growing knowing:

1. Are each unique in and of themselves due to their being in the process of acquiring unique experiencing as they move through time and space.

   Each entity of knowing has its own unique perceptions and experiencing assembled uniquely by the linear progression of the multiple facets of both time and space or whatever abstractual fabric should exist within the 'universe' from which the entity of knowing emerges

2. Vary in size since relative to each other they contain vastly different 'quantities' of knowledge and contain vastly different experiences.

3. Not only may appear but do vary regarding their 'distance' apart since they are immersed in a location whose very fabric is composed of distance. In essence, the 'distance' separating them does exist.

   They are therefore all separated by varying degrees of distance no two distances of which are equal

4. Are dependent upon beginning – end parameters since they are immersed in time and space, are immersed within a form of Centrism

Process

5. Take 'time' to get from one to another, to get from one place to another, since there is 'distance' to traverse

6. Are incapable of 'knowing' one another completely since they are not complete forms of knowledge but rather are actively and uniquely experiencing knowledge and being assembled through the active process of knowing and experiencing uniquely.

   The process itself becomes uniquely an integral part of themselves since the active process of gaining knowledge is the creation of the perception of knowledge and experiencing through the process of knowing itself.

Potentiality

7. Are incapable of incorporating other 'newly' developing entities of unique knowledge, knowing, and experiencing for their process of 'forming' has not yet ended and therefore any new knowledge and experiencing 'affects' the very formation of the entity itself is in the process of 'becoming'.

8. Partial summations of a unique entity of knowing as well as the complete summation of the unique entity of knowing is as varied as the potential combinations the parts of 'a' particular unique entity of developing knowing can allow confined to itself. This is known as 'limited' knowing.

9. Potentiality within region #1 is, due to #1 – 8, limited to the abstractual and ? and ? fabric of a particular universe. In 'our' case, potentiality regarding formation of unique entities of knowing is limited to what time, space, matter, and energy will allow.

Since uniqueness is a quality acquired by the active process of knowing itself, each entity of knowing is unique and becomes so independent of one another

Independence thus becomes the principle of region #1.

This is not to say that individuals are not dependent upon one another in society. What it says is that the very concept of uniqueness is a quality of existence itself and is not a characteristic 'given' by one individual to another.

Centrism is not a difficult concept to comprehend because we are, after all, immersed 'within' Centrism', and as such we not only observe but experience time and space to be concepts expanding 'outward' from 'a' point of reference which varies from conscious knowing to conscious knowing.

What must not be lost within the exploration of the familiar, however, is that within a location whose very fabric is that of space and time, 'correct' sequencing is a fundamental principle.

One cannot die unless one is first born, one cannot wake up unless one first goes to sleep, and one cannot swim unless one first goes into the water.

Granted the term 'cannot' may be too strong a term to use in these particular examples but the concept of 'correct' sequencing is an aspect of time and space and as such is something all of us within our personal universe understand.

Region #1 therefore is a location where our unique experiencing does not depend upon what others have to offer us to experience.

Rather our own experiencing is in the process of 'becoming' as uniquely experienced by ourselves.

Our own unique experiencing and knowledge as viewed uniquely through our own knowing is becoming what it is - our own uniqueness.

We are in the process of becoming rather than being what we finally are to be.

Panentheism
Addressing
Anthropocentrism

We can change with the infusion, embracing, union of experiencing space and time while immersed within matter and energy.

Divine intervention? Perhaps. Who is to say what the whole is capable of doing within its creation: the universe, our reality.

How can an 'all knowing' entity be all knowing if it doesn't know 'all things'?

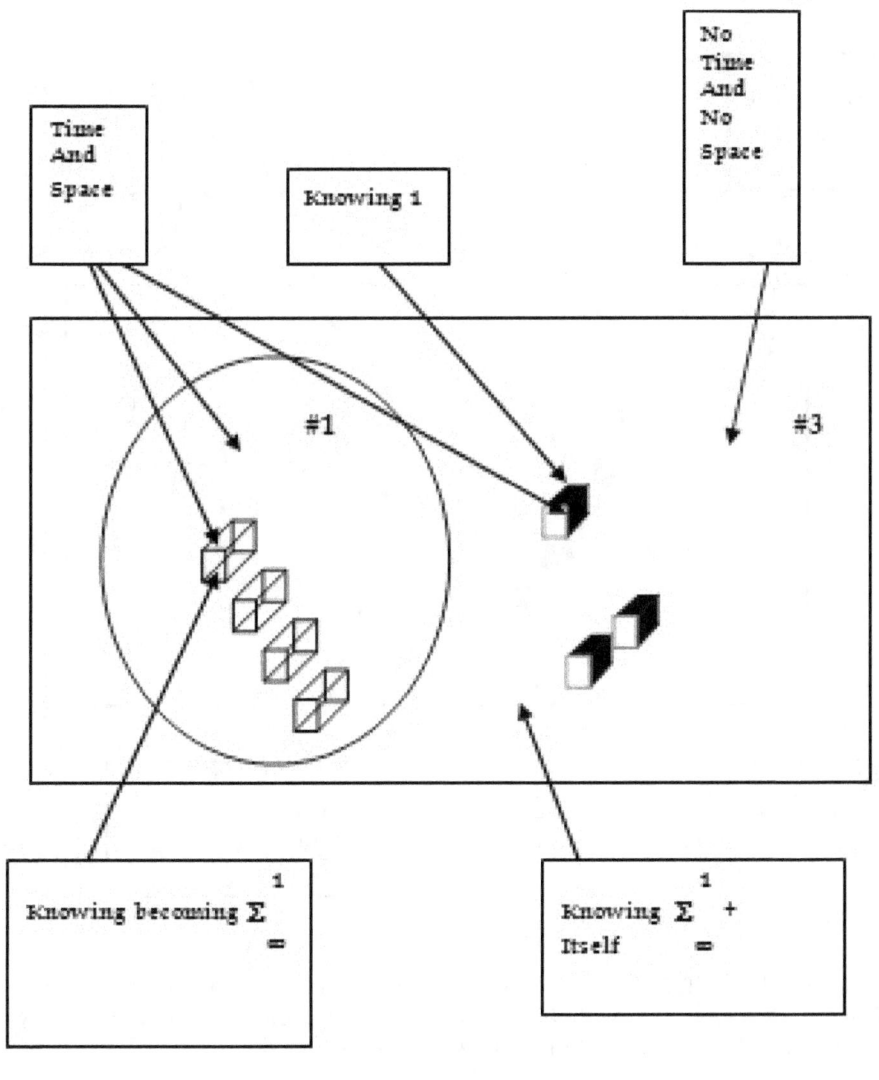

Again we must ask: Doesn't this diagram imply 'all Knowing' doesn't know what 'will be' and doesn't this in turn imply time exists in 'all Knowing'?

Again, we must reply: The concept of what 'will be' only exist 'within' time. If we review what we had previously learned of time we find time to be 'located' within two locations:

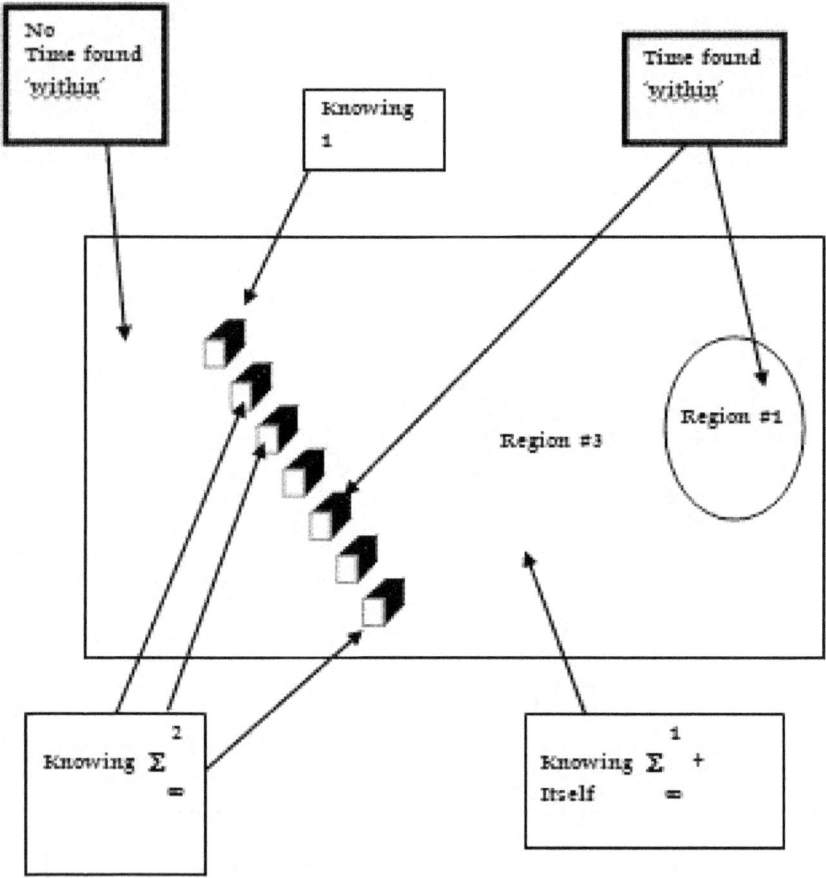

Since time is not found either as an innate characteristic of region #3 nor as a medium of region #3 'within' which subsets of region #3 find themselves immersed, there is no concept of 'what will be' to be found in region #3.

Panentheism
Addressing
Anthropocentrism

This is not to say time is not found 'within' region #3, rather it demonstrated time is not an all-pervasive characteristic, a universal medium of region #3.

Time is found in region #3 in two distinct location of which we are presently capable of perceiving.

Time is found 'within' entities of 'knowing', which have evolved out of the 'universe', and time is found as a medium of the universe within which entities of 'knowing' move from being 'virgin' entities of 'knowing' to being complete entities of 'knowing'.

Where then does the 'Book of Divine Knowledge' find itself to be in terms of the metaphysical system of singular location, the individual 'acting within'/being a part God?

Daniel J Shepard
Channel

Panentheism
Addressing
Anthropocentrism

17. The significance of insignificance: Random Sequencing

There is really no place to begin with such a topic. Perhaps that is a fitting observation considering the topic itself: randomness.

We have, however, little choice but to begin if we are to understand the significance of insignificance.

So lets begin:

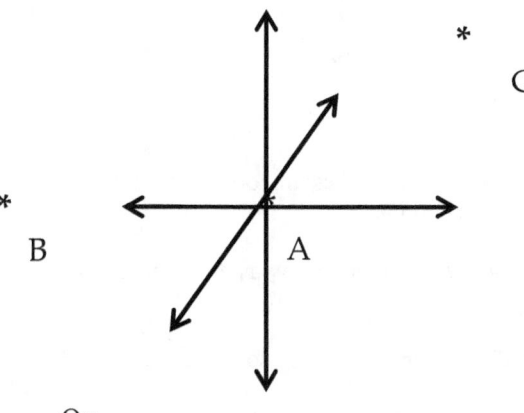

Or:

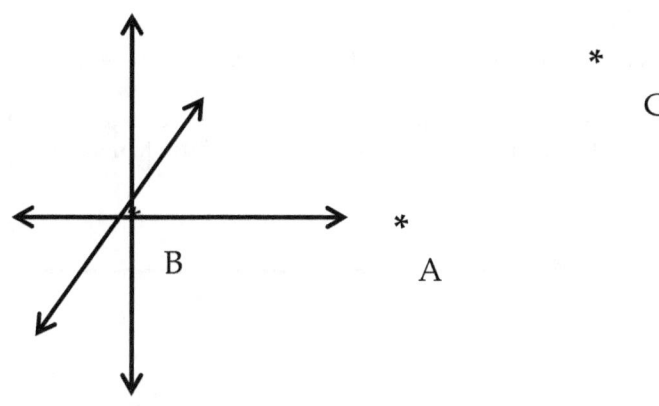

Or:

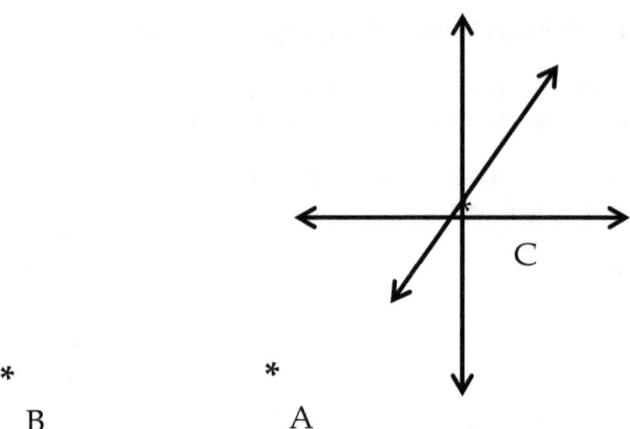

The potential names of points A, B, and C using the Cartesian coordinates (X, Y, Z) grows as the number of locations for the origin increases.

The number of locations of the origin grows exponentially as the number of dimensions grows linearly.

One could begin with one point and zero dimensions. By doing so one observes

*
A

Since a point has no length, depth, or height, point A has zero dimensions.

By adding one dimension we obtain: zero dimensions + one dimension = one dimension but the potential names for 'a' point becomes infinite in nature:

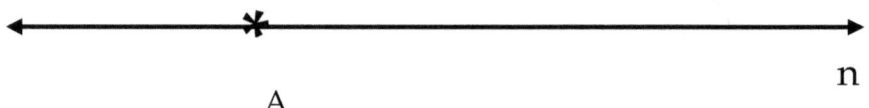

Panentheism
Addressing
Anthropocentrism

The point A has the potential to be located anywhere upon line n.

The concept of one dimension being the only existence creates a line, line n, which is infinite in length.

Point A has the potential of being located anywhere upon line n since, by definition, point A has no length.

Point A could move along line n, however, point A need not move to change position relative to line n, line n could move relative to dimensions 2, 3, 4 ... which exist to us but not to the situation we are discussing.

Such a statement is neither a paradox nor an untruth.

As we, you and I can readily attest to, just because 'a' dimension and its components are not aware of other dimensions does not mean other dimensions do not exist.

If we now add a second dimension, we can observe the nonlinear growth of potential points of location for point A.

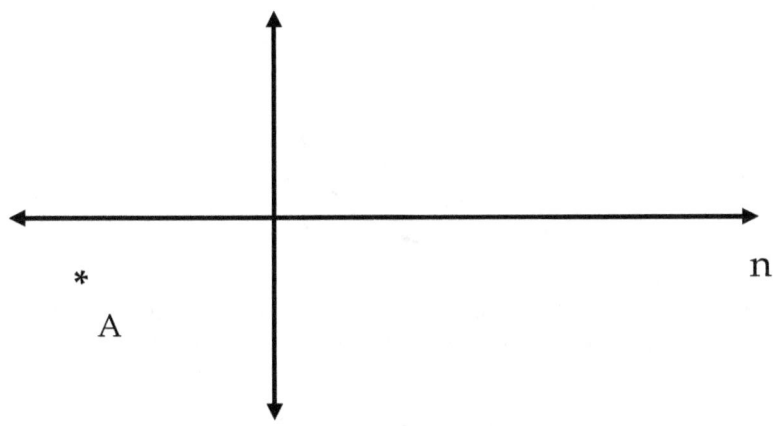

So it is two and then three dimensions can be understood to exponentially grow the perceptual potential coordinates for point A.

If we presume point A can move or if point

Daniel J Shepard

Channel

A is presumed to be static in its location the same results of exponential growth for the names of point A can be understood to increase if we presume the X and Y axis move.

If we then factor in the name of point A changing with time, again the perception regarding point A's contribution to the space/time relationship causes another exponential increase in the coordinate names point A could acquire.

But what does this have to do with abstractual concepts of free will, determinism, significance and insignificance?

This progression of thought regarding the exponential expansion of perceptual spatial location generated through increasing numbers of dimensions leads us to the rudimentary mindset which will help us understand the exponential expansion of the totality of perceptual 'knowing' generated through increasing numbers of unique entities of 'knowing' evolving out of free will.

To understand the connection let us look at a unit entity of awareness, a unit entity of 'knowing' as it develops within the physical:

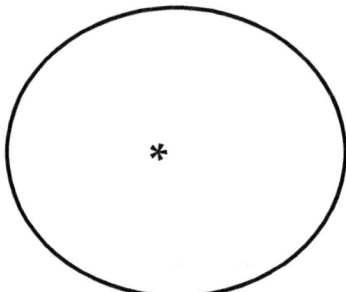

Virgin awareness, virgin 'knowing', begins with no 'knowing' and expands through experiencing within the universe. As such, it could be drawn as:

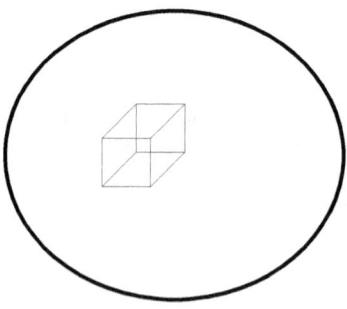

Panentheism
Addressing
Anthropocentrism

The transparency of the rectangular prism represents the ability of the physical to continue to affect and form the entity of 'knowing'.

The entity of knowing continues to grow, experience, and formulate its completeness of unique 'knowing 'until it' dies.

At the point of it's no longer being capable of continuing to grow its summation of awareness; we obtain what could be diagramed as:

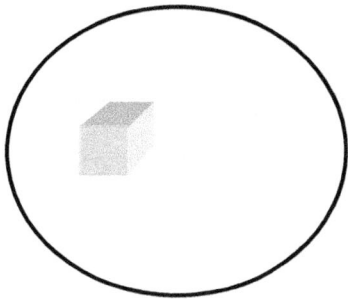

The ability to initiate actions of free will combined with the ability to experience actions bound by the laws of nature generate the unit of unique abstractual perception which has developed.

Now lets remove the influence of the physical since, as symbolized by the 'filled' rectangular prism, the sensory mechanism has been removed from the individual unit of perceptual development.

This removal of the physical does not imply the abstractual uniqueness of the unit of 'knowing' just 'goes away'.

In fact, the implication of the individual 'acting within'/being a part God implies quite the contrary. As such, we obtain the following:

   Unit A

We have removed physical reality in order to study the entity of awareness.

We will label this entity: unit A. Keep in mind.

Unit A is no longer a Virgin point of 'knowing' but rather a point of 'knowing' having its own unique perceptual 'outlook' which has been developed through the influence of an almost infinite number of abstractual interactions.

Such interactions have been generated through the actions of both free will initiated by itself and initiated by actions of free will generated by other units of 'knowing', as well as generated by actions bound by the laws of nature.

In essence, the unit of abstraction indicated above is 'filled' with various perceptions, desires, wants, loves jealousy, greed, compassion, etc.

One must also recognize the unit to be just that: 'a unit'. It is. It is unique. It has 'wholeness' of perception that is unique in and of itself.

Before we move any further, one will find it interesting to note that if Unit A is the whole of existence, then:

1. The potential perceptual development of the whole is simply that of unit A
2. The number of abstractual existences are infinite in number but limited to what is found in unit A
3. Perceptions of time and distance can be found 'within' unit A but not 'outside' unit A for there is nothing 'outside' unit A
4. The 'whole' is identical to unit A and as such has the same perceptions as unit A
5. The 'whole' has the power of unit A and no more
6. The 'whole' has the same 'knowing' as unit A and no more
7. The 'whole' has the same 'presence' as unit A and no more

Sound familiar? The discussion evolves out of Tractates 1, 2, and 3: Zeno, Aristotle, and Boethius.

In order to address our understanding of what we mean by the whole we will enclose unit A within the whole:

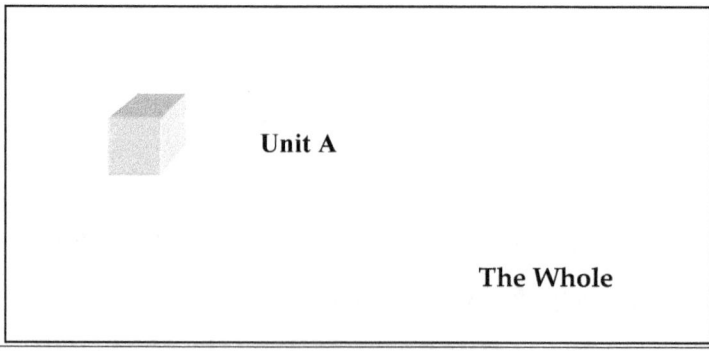

Panentheism
Addressing
Anthropocentrism

It is obvious the whole is Unit A.

It is also obvious time and distances are abstractions found 'within' unit A for the elements necessary to generate the concepts of time and distances are no longer present.

Time and distance are elements of space and matter both of which had been removed when we erased the circle that represented the physical reality of the universe.

What then happens if a second unit of 'knowing' is added to the system above?

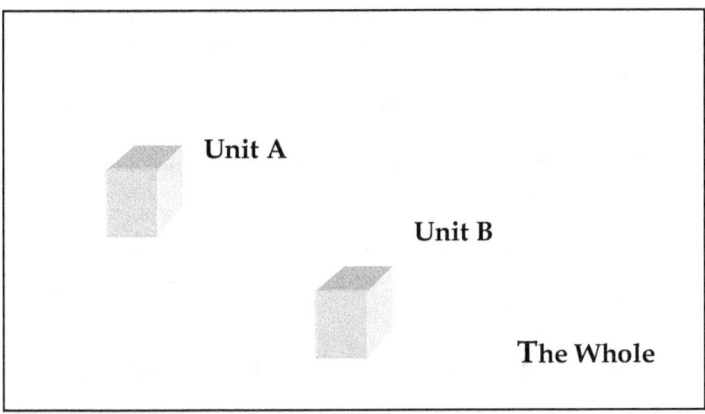

Now we no longer can say:

1. The potential perceptual development of the whole is simply that of unit A
2. The number of abstractual existences are infinite in number but limited to what is found in unit A
3. Perceptions of time and distance can be found 'within' unit A but not 'outside' unit A for there is nothing 'outside' unit A
4. The 'whole' is identical to unit A and as such has the same perceptions as unit A
5. The 'whole' has the power of unit A and no more
6. The 'whole' has the same 'knowing' as unit A and no more
7. The 'whole' has the same 'presence' as unit A and no more

We must now say:

1. The potential perceptual development of the whole is that of unit A, unit B, or unit A + unit B
2. The number of abstractual existences are infinite in number but is no longer limited to what is found in unit A, rather it is limited to what is found in unit A, unit B, or unit A + unit B
3. Perceptions of time and distance can be found 'within' unit A or found 'within' unit B but not 'outside' unit A and/or unit 'B' for there is nothing 'outside' unit A and unit B
4. There is now something found 'outside' unit A and that is not only 'unit' B but the summation of unit A + unit B
5. The 'whole' is identical to unit A, or unit B, or unit A + unit B and as such has the same perceptions as unit A or unit B or unit A + unit B
6. The 'whole' has the power of unit A or unit B or unit (A + B) and no more
7. The 'whole' has the same 'knowing' as unit A or unit B or unit (A + B) and no more
8. The 'whole' has the same 'presence' as unit A or unit B or unit (A + B) and no more

It becomes apparent that the increase in the varieties of combinations of the whole increase not on a linear basis but rather on some forms of geometrical or exponential basis.

If we expand our units of unique 'knowing' developed under the laws of free will, we obtain three units of unique knowing and as such obtain:

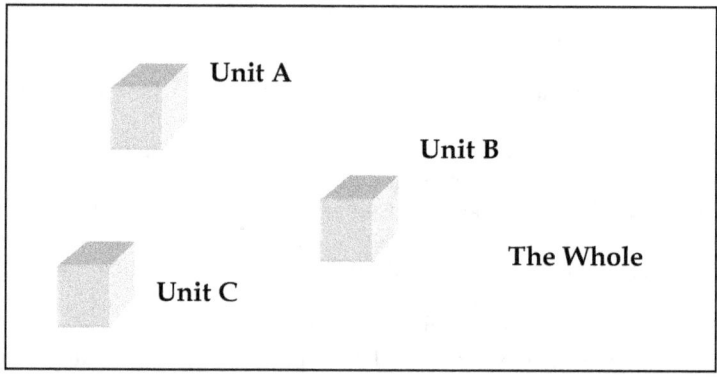

At this point, we could reevaluate the results of the above using the same format we previously used.

This process however becomes beset with even more verbiage than previously.

Panentheism
Addressing
Anthropocentrism

To minimize this problem of verbiage we will examine the results of the above using more diagrams.

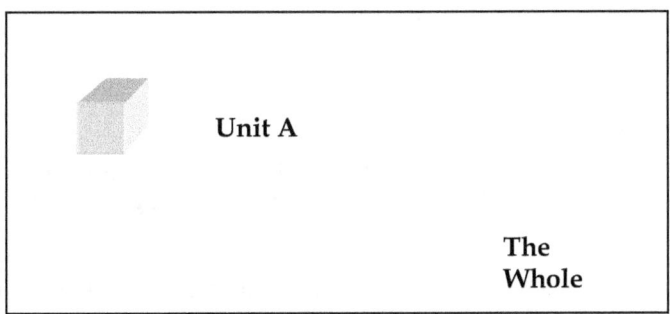

We will now open unit A and pour its 'substance' into the Whole.

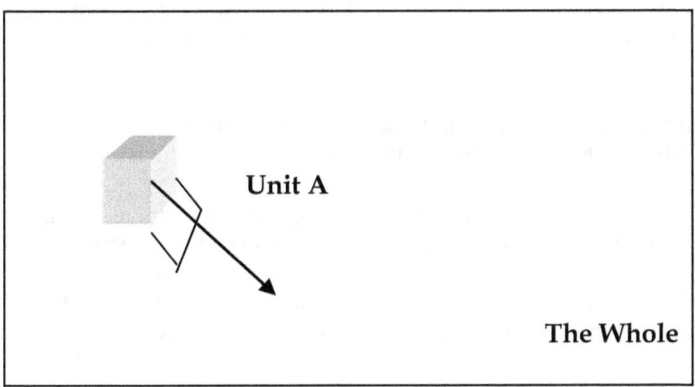

Example 1: If Unit A only

    Possibilities for the Whole:

        1. A

Example 2: If Unit A and Unit B only

Possibilities for the Whole:

    1. A
    2. B
    3. AB
    4. BA

Before we go to example three, we should address the question of why the potential for BA? AB and BA are not the same even though time and/or distance are not 'active' elements of the whole.

Time and distance are elements of Unit A and Unit B. As such, should the contents of A and B be released 'into' the Whole, then time and distance become options with which the Whole can develop its own unique perceptions.

At first glance, it would appear the characteristic of cardinality would only be relevant if one speaks in terms of sequencing physical perceptions generated through the element of cardinality innately found in spatial cardinal concepts of distance or if one speaks in terms of sequencing abstractual perceptions generated through the element of cardinality innately found in the abstractual cardinal concept of time.

The perceptions generated through the potential combinations of A and B are not A, B, and AB but also BA. A fifth state of the whole itself exists.

This fifth state of being is the summation of some sequential form of A and B independent of the perception developed through an order of sequencing.

Having said this we then obtain a new perspective for the possibilities for the whole, which in turn will allow us to move to example three.

Example 2 corrected:

    Possibilities for the whole:

        1. A
        2. B
        3. AB
        4. BA
        5. the whole

Panentheism
Addressing
Anthropocentrism

Example 3:

    Possibilities for the whole:

1. A
2. B
3. C
4. AB
5. BA
6. BC
7. the whole of A and B
8. ABC
9. ACB
10. BAC
11. BCA
12. CAB
13. CBA
14. the whole of A and B and C

There is a pattern developing here.

The mathematics of it are much too difficult to explore in this tractate.

However, a simplistic example can be demonstrated by following a similar yet much more simplistic sequence:

Example 1:

    A

Example 2:

    A, B, AB

Example 3:

    A, B, C, AB, AC, BC, ABC

Example 4:

    A, B, C, D, AB, AC, AD, BC, BD, CD, ABC, ABD, ACD, BCD, ABCD

As one examines the increase in potential possibilities generated by the sequence one begins to appreciate the significance of each unit.

Daniel J Shepard

Channel

If there is only one unit, the unit is the whole, there is nothing greater.

If 'a' second unit is added the number of potential possibilities expands by two not one, expands to three.

If 'a' third unit is added the number of potential possibilities expands by four not one, expands to seven.

If 'a' fourth unit is added the number of potential possibilities expands by eight not one, expands to fifteen.

Any one unit of growth may or may not be greater than another.

The size of the unit is not what is important here but rather what is important is the very fact that an additional unit has been added.

Mathematically we see the pattern as being:

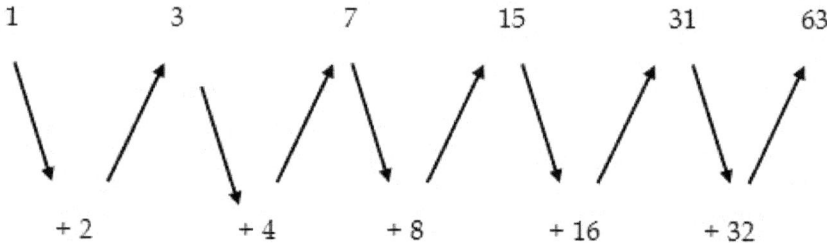

As one can see, the potential increases non-linearly.

Each unit added has tremendous repercussion upon the potential of the whole.

Each unit added has tremendous significance to the whole.

Each unit added impacts the whole more than once, more than itself.

But where is the individual unit here? It appears to be lost.

Panentheism
Addressing
Anthropocentrism

It appears to be lost not because it is lost but because we have not fully expanded upon the pattern.

If we more fully expand the pattern to be what it is, we obtain:

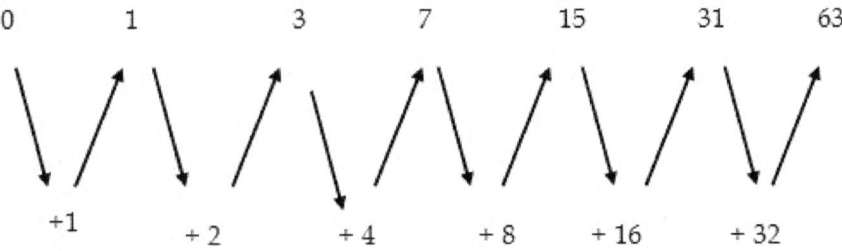

This does not appear to fit the pattern unless we adjust the pattern accordingly:

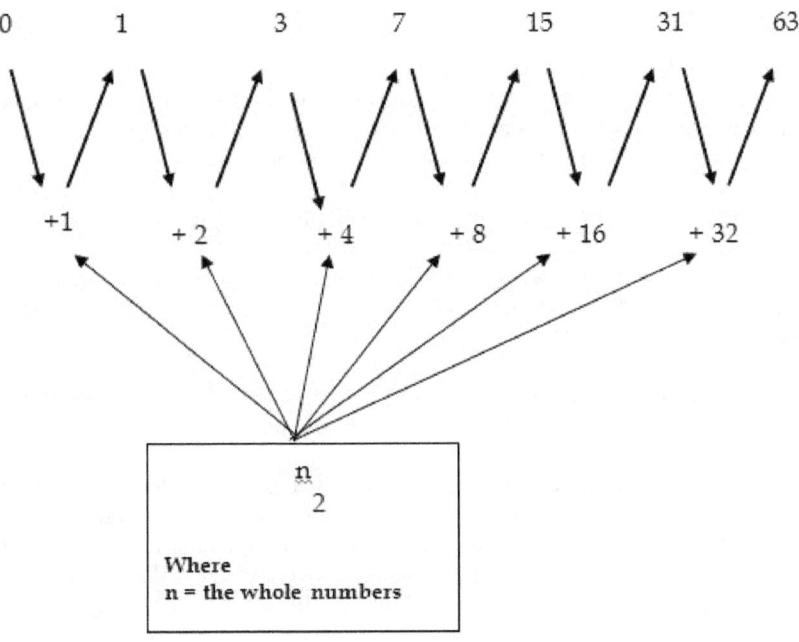

All of a sudden, we see the base for what it is.

Daniel J Shepard
Channel

The base, the foundation, is built upon the individual unit since 'n' is the epitome of individuality itself. 'n' is the set of unit numbers in their very completeness of form. 'n' is the set of natural numbers beginning with one and extending into infinity as a set comprised of increasingly large numbers of whole, complete units based upon 'a' whole unit itself.

As we can see, the growth is not geometric but exponential.

Is this an anomaly or is this the very concept upon which the growth of the whole itself is built when one speaks of 'all knowing'?

Within the system of the individual 'acting within'/being a part God, this is not an anomaly but rather the foundation of total knowing, the whole itself.

Under the metaphysical system of panentheism, the individual 'acting within'/being a part God, the very perceptual abilities of the whole grow exponentially with each unit of knowing added to it. Furthermore, the growth is not based simply upon the pattern diagrammed above but rather the pattern leaps beyond this potential by a factor of:

The whole of A + the whole of A and B + the whole of A and B and C + ...

So it is each individual unit added may be added to such a large number of others that it appears insignificant compared to the whole but it is in fact 'the' element which increases the whole exponentially over the huge 'potentiality of 'knowing'' it had been previous to the individual's addition to the whole.

Doesn't the concept of 'previously' imply an element of time and sequencing?

No and that is the very reason the concept such as alphabetical sequencing, a concept of sequencing 'controlled' by time becomes a limit not imposed upon the whole of abstraction itself.

The result of course is an even greater form of exponential growth applying to the whole than the form of exponential growth we demonstrated.

Suddenly, through an understanding of a new metaphysical system – through an understanding of a non-Cartesian system powered by a Cartesian system – through an understanding of the individual 'acting within'/being a part God, through an understanding of a system of determinism powered by free will, we begin a new understanding regarding the very significance of the seemingly insignificant.

Humanity, the individual, begins its upward climb out of the depths of insignificance and into the glory of significance itself.

Panentheism
Addressing
Anthropocentrism

18. The explosive nature of the potentiality of knowing

Adding 'a' piece of knowing creates an explosion of potential combinations over and above what existed previously.

As each new piece of knowing is added, the addition creates an exponential expansion of potentiality to which the next piece of knowing can add its potentiality of growth to the whole.

Potentiality now becomes a situation of 'expanding', increasing, potential growth itself on an exponential basis as opposed to an 'expanding', increasing potential growth on a geometrical basis.

The whole itself now gains not just the potential to grow but gains the potential to grow in a potentially explosive manner.

The difference may best be understood as that of the difference in existence of a substance, the explosive nature of the substance gun powder and the explosive nature of fission, and the explosive nature of fusion and now: the explosive nature of knowing.

What does this have to do with Centrism and non-Centrism?

It is the very concept of Centrism, which limits the potentiality of the Whole itself. It is Centrism that limits our potential significance.

And the further away we appear to be from the center of the whole, the less significant we perceive our significance itself to be.

This was not the intent of Copernicus as he reevaluated the concept regarding the 'location' of the center.

As little as Copernicus had expected to influence our very understanding regarding the significance of the individual, his work involving the search for the center of the physical was to impact humanity's most fundamental perception regarding the value of the individual and its own species.

Daniel J Shepard
Channel

Panentheism
Addressing
Anthropocentrism

19. Removing a piece of Randomness

To understand the significance regarding the impact removing a piece of knowing from the whole has upon the whole itself, one must not begin by removing what one find at the left but at the right of the graphic:

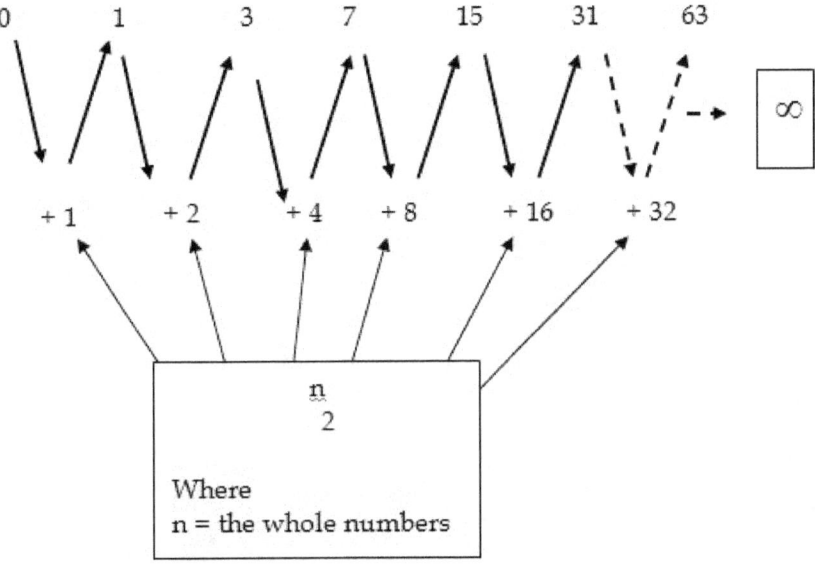

And with that the significance of removing, losing, terminating, interfering with, 'cutting short', a piece of knowing's potential once it has 'become', speaks for itself. Interfering with the development of a unit of knowing affects the outcome of what the Whole is.

Daniel J Shepard
Channel

Panentheism
Addressing
Anthropocentrism

20. Boethius' metaphysical system and why we can now file it away as a part of the annals of history

We must now come back to the diagram representing the 'location' of free will and the 'location' of determinism in order to understand why it is we can finally relegate Boethius' metaphysical system to the annals of history, relegate Boethius' metaphysical system to that of being a history book as opposed to being a current philosophical theme.

This aspect of Volume 7: Boethius, was put on hold until we were able to expand our understanding regarding Centrism and its affect upon units of knowing addressed within this tractate.

We can now return to the concept of free will and determinism for a short summation as to why Boethius' metaphysical system can now be filed away in the archives of interesting historical paradoxes.

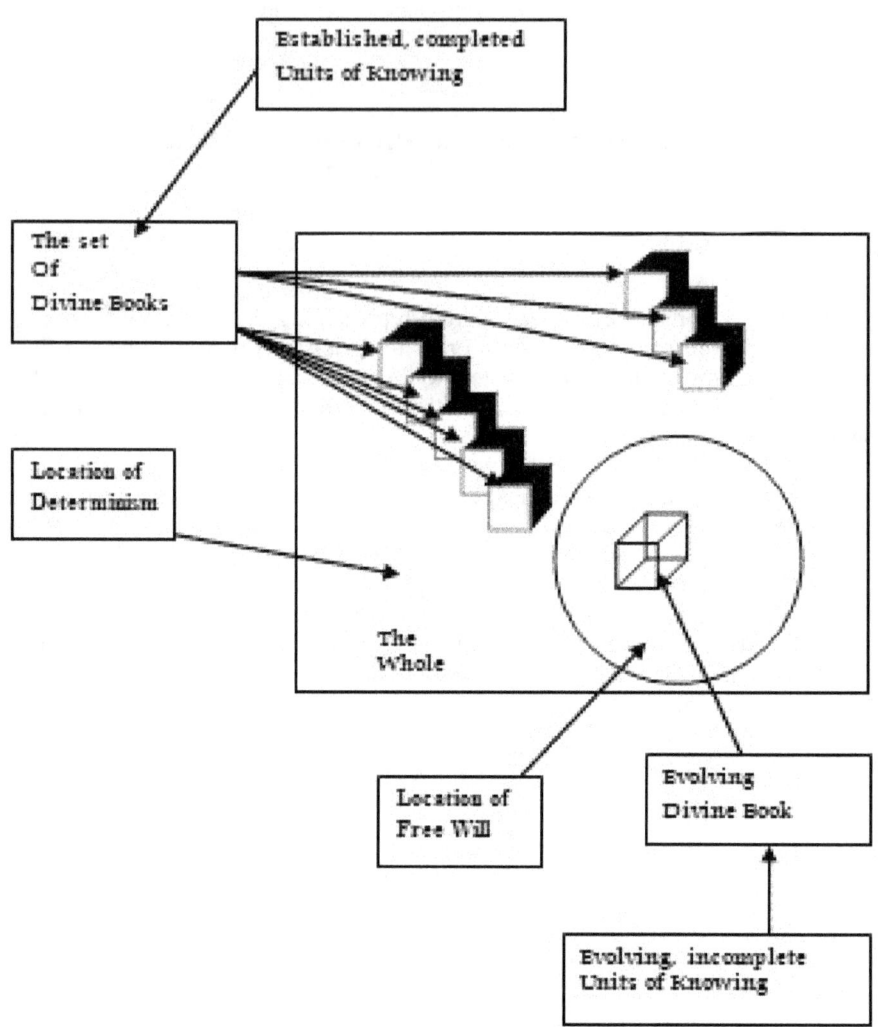

As we have done so often before, we find ourselves in need of simplifying the diagram.

Panentheism
Addressing
Anthropocentrism

Simplifying we obtain:

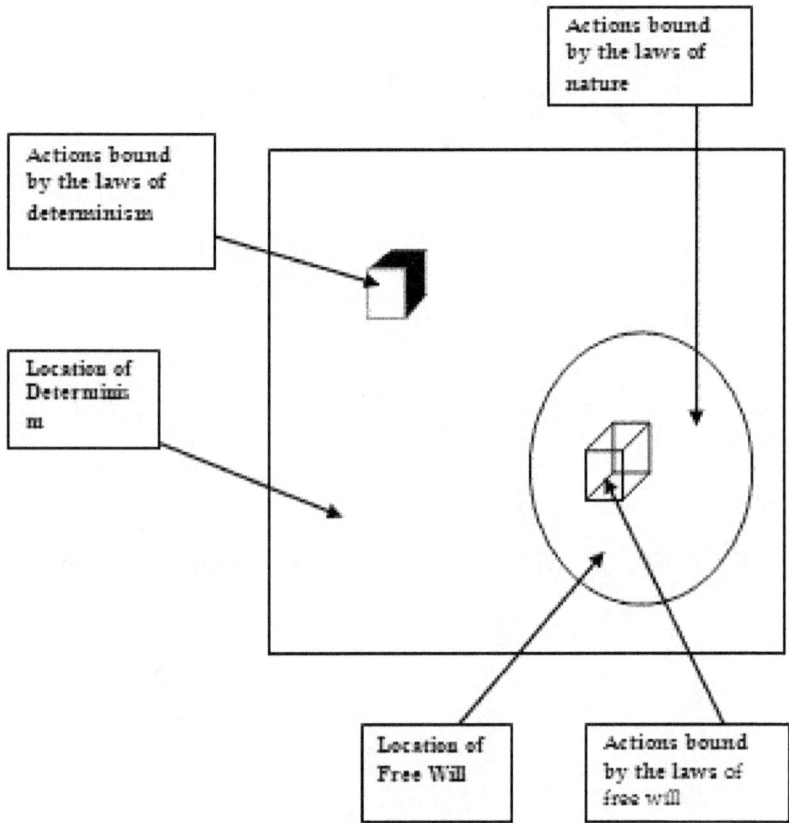

New questions now begin to emerge. Questions emerge which had no opportunity to emerge under the confines of Boethius' system where determinism was bound 'within' the same confines as free will.

Is the whole confined by actions bound by the laws of free will? Graphically we now understand such a question to be represented as:

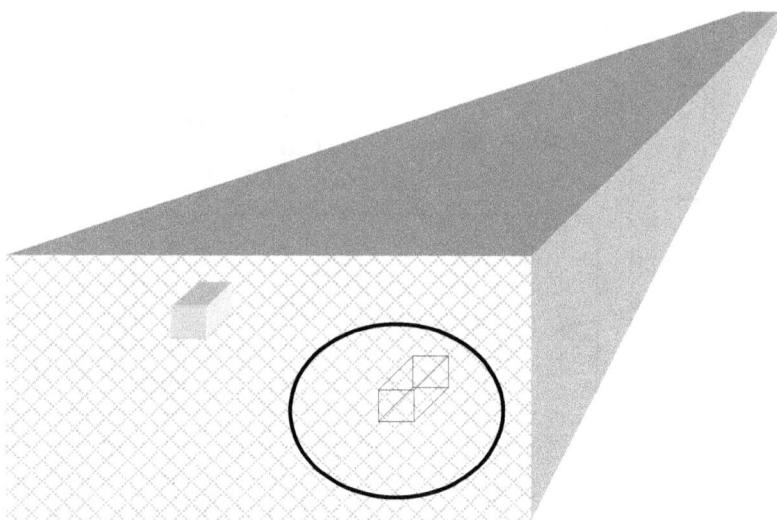

Is the whole confined by action bound by the laws of determinism? Graphically we now understand such a question to be represented as

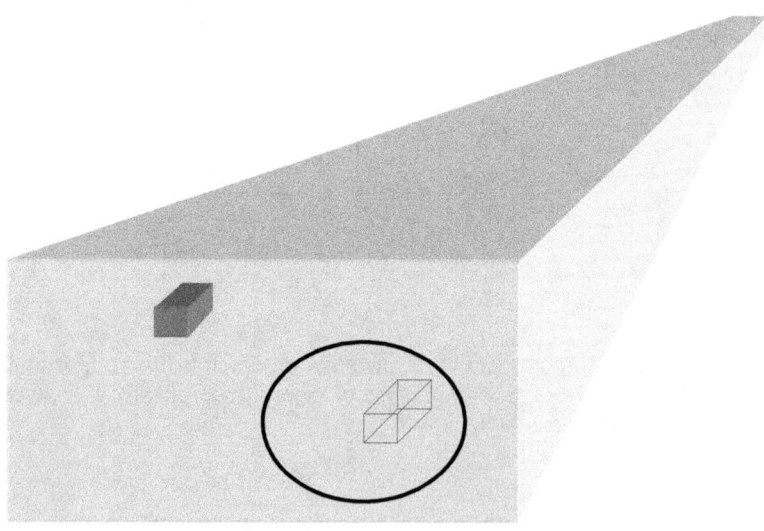

Panentheism
Addressing
Anthropocentrism

Is determinism found 'within' the location of free will? Graphically we now

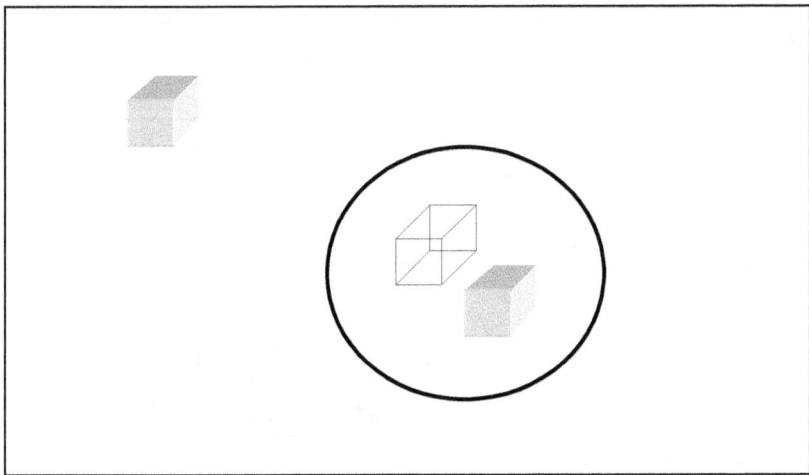

Is free will found 'within' the location of determinism? Graphically we now understand such a question to be represented as:

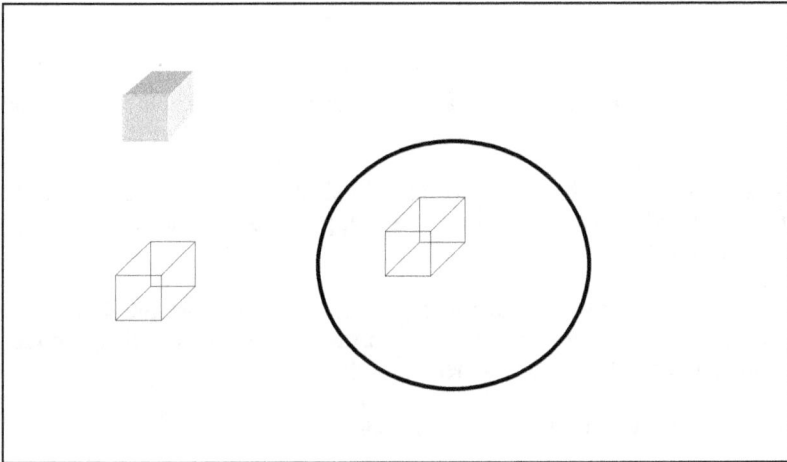

Etc., etc., etc.

Daniel J Shepard
Channel

The questions, not the answers, have only been presented.

The questions are limitless, the answers now become intuitively understandable through the application of the new metaphysical perception: the individual 'acting within'/being a part God or generically speaking 'panentheism.'

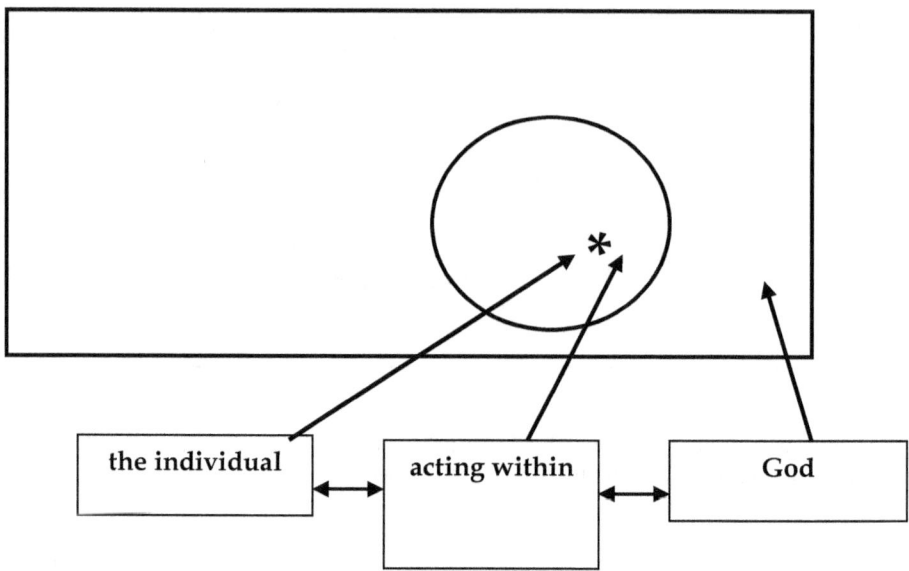

Some of these questions are not metaphysical in nature but rather are ontological in nature.

It is for this very reason that this new metaphysical perception of a non-Cartesian system being powered by a Cartesian system appears to relegate Boethius' system to the annals of history.

Boethius' system appeared to leave metaphysics in a state of stagnation while this new perception opens up an almost infinite array of new metaphysical thoughts as well as philosophical thoughts in general.

But what of the answers to the given questions?

How can one possible bring up such topics of discussion and then leave the reader in a state of suspended anticipation?

Panentheism
Addressing
Anthropocentrism

For the time being, we have little choice but to proceed with this work regarding The War and Peace of a New Metaphysical Perception.

However, if it is of any solace, these questions will come up again in Volume 8:

The End of the Beginning – Theoretical Metaphysics Emerges.

For the time being, however, we must get back on track or we will never get done with the examination of many of philosophy's greatest paradoxes and how it is we can now resolve them and as such relegate not just Boethius' paradox but a large number of paradoxes to the annals of history.

Daniel J Shepard
Channel

Panentheism
Addressing
Anthropocentrism

21. Archimedean Points

Archimedean Points are referred to by Husserl as 'the' Archimedean Point. The Archimedean Point is the unshakable foundation of human knowledge.[xi]

It is 'within' the individual, be it the subset of the whole as 'the individual' or 'the whole' as the individual, 'within' which, the 'unit' of knowledge is found.

In addition, it is 'within' the individual, be it the subset of the whole as 'the individual' or 'the whole' as the individual, 'within' which, the process of knowing knowledge is found.

The concept of the individual 'acting within'/being a part God now becomes two 'substantives', or two universals, as Russell would say, interacting upon each other via the verb.

The universals, substantives: the individual and God

The verb: 'acting within'/being a part

All three are elements of the system.

To rephrase it, we have a system of: three in one.

So it is:

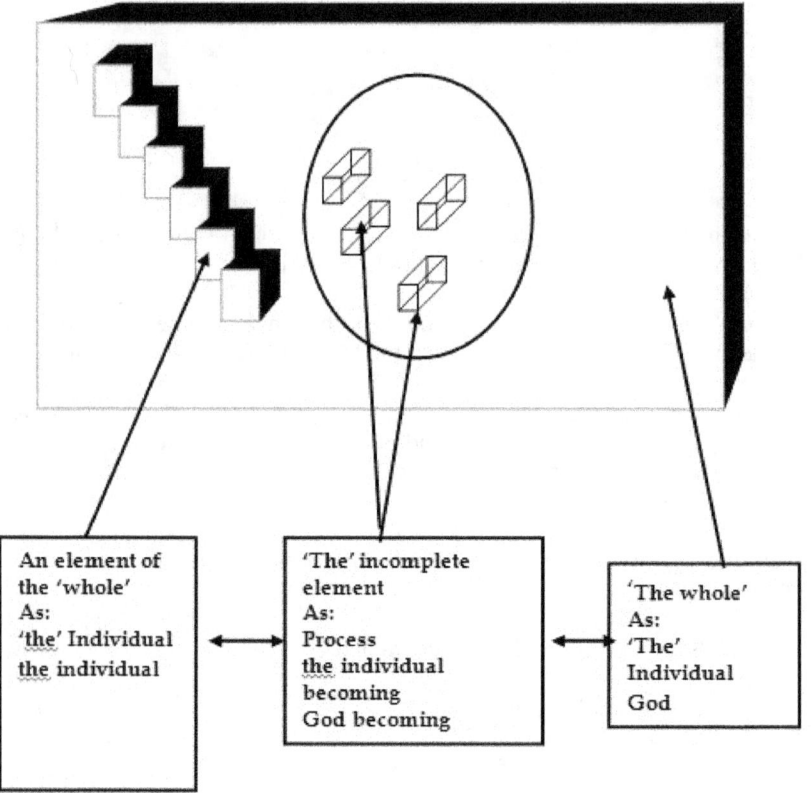

So it is each entity of knowing becomes 'a' first truth in relationship to itself, in relationship to the whole, and in relationship to the universe/reality

So it is 'infinite relative 1st truths' emerge within 'finite relative 1st truths' within 'a' 1st truth.

Panentheism
Addressing
Anthropocentrism

22. Philosophical infinities

Juxtaposition of infinities thus emerges as an essential element of the new metaphysical system the individual 'acting within'/being a part God.

The concurrence of ontological infinities, cosmological infinities, and metaphysical infinities arises out of the understanding of this new metaphysical perception.

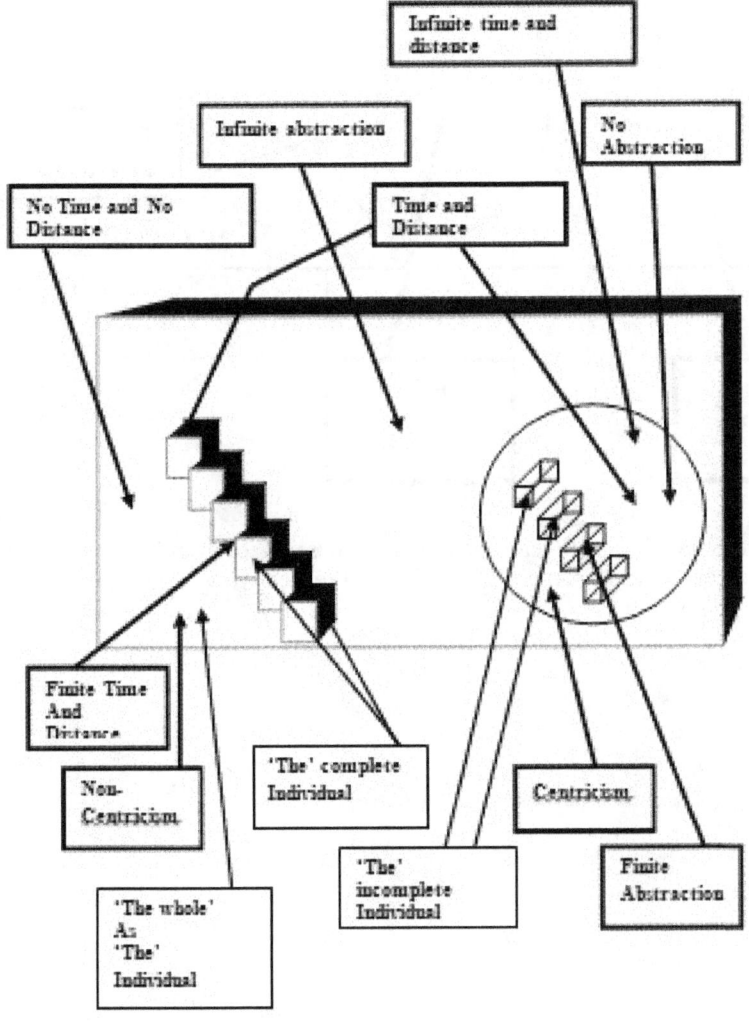

We can simplify the above diagram as:

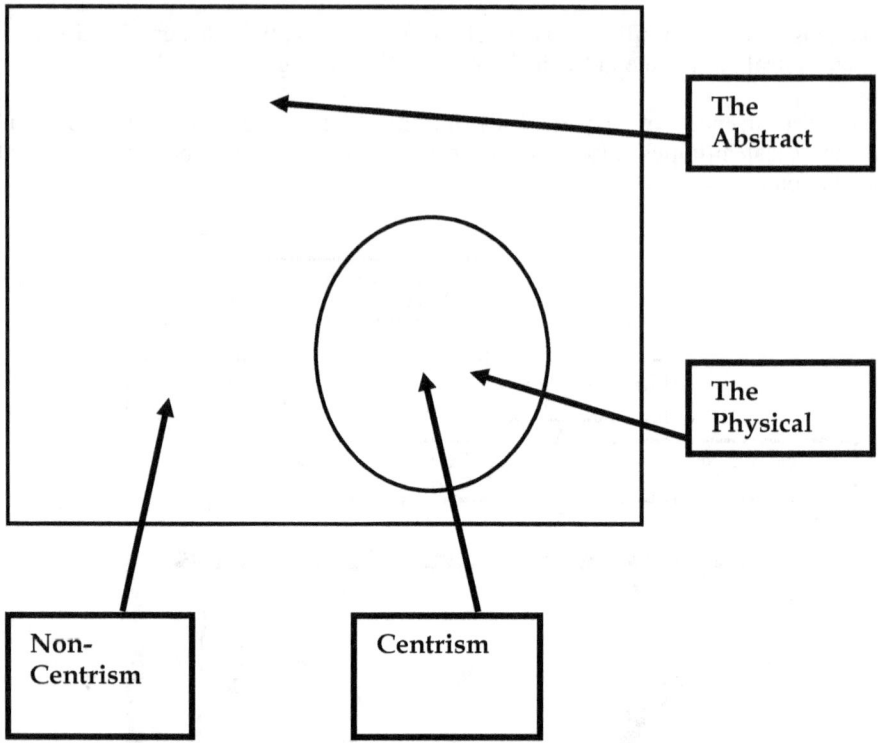

Such an understanding will lead us into a discussion of Kant's Centrism and Hegel's non-Centrism in Tractates Six and Seven respectively.

Panentheism
Addressing
Anthropocentrism

The latter diagram evolves through the process of reality/the physical itself being 'experienced' through a process, which can be depicted as:

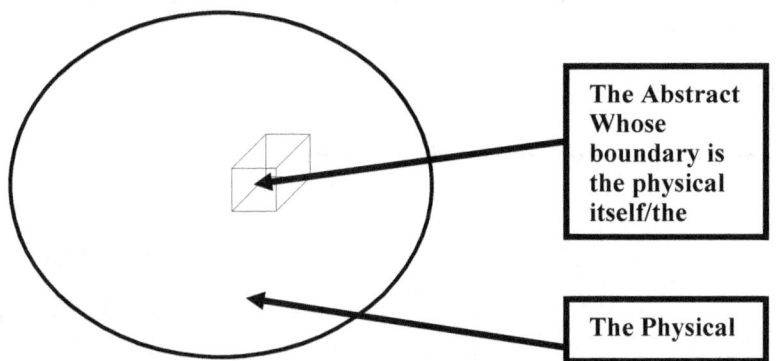

Which, when placed back into its former diagram, becomes:

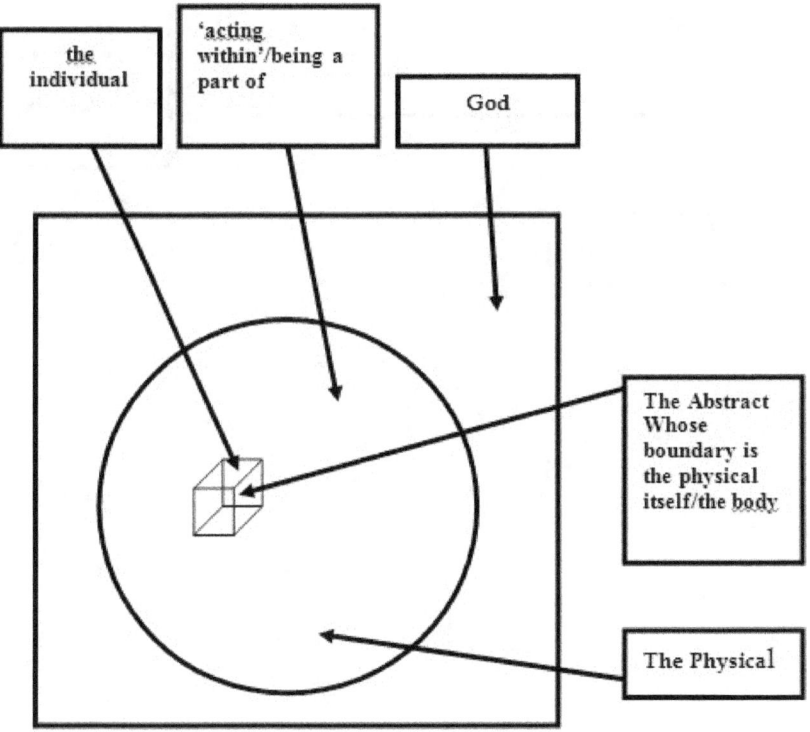

Or the individual 'acting within'/being a part God, Centrism within non-Centrism, Cartesian within non-Cartesian, non-Cartesian powered by Cartesian, non-Centrism powered by Centrism, God powered by the individual through 'acting within'/being a part of God.

This in turn becomes:

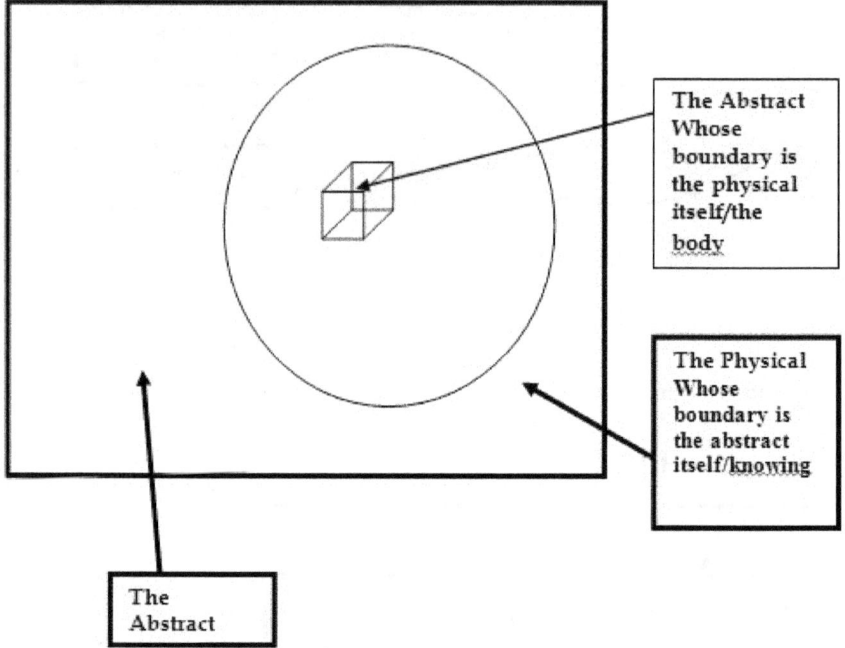

Panentheism
Addressing
Anthropocentrism

Or to put it more generically:

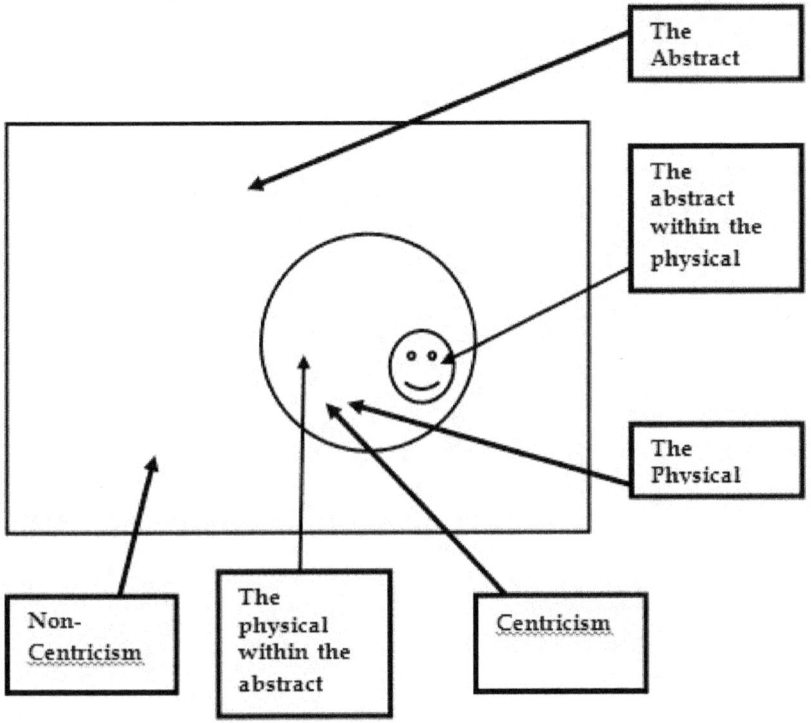

Daniel J Shepard
Channel

Panentheism
Addressing
Anthropocentrism

23. A bag of marbles is not dependent upon sequential time

Remove the universe, remove the physical and we have:

Daniel J Shepard
Channel

Round out the unit of knowing:

Duplicate the units of knowing:

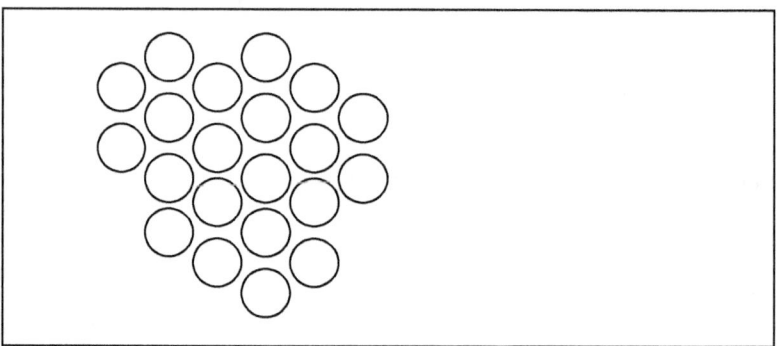

Resize your container

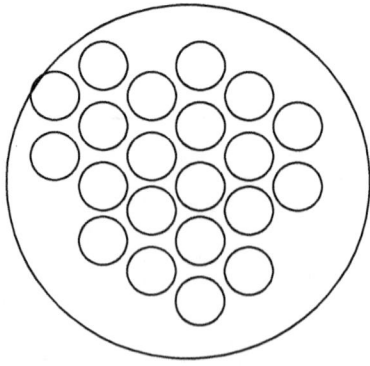

Panentheism
Addressing
Anthropocentrism

And you have a bag of twenty-one marbles

So what?

So, the twenty-one marbles are not dependent upon time for their existence.

If you put your hand in the bag and mix the marbles, you may have rearranged the marbles but you still have twenty-one marbles.

Not only do you have twenty-one marbles but each marble is as it was as opposed to is where it was.

Now lets place time into the picture.

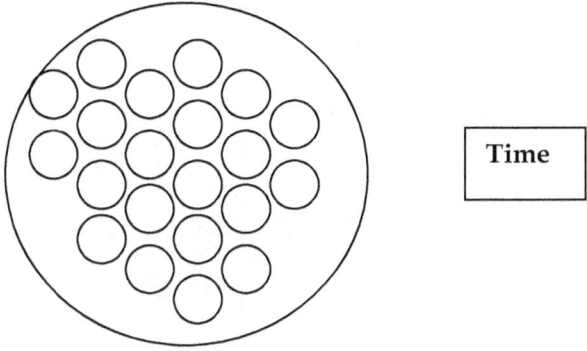

The question becomes: Where does time fit into the graphic?

The marbles went through a process of emerging as individual entities, unique individual entities.

Each marble emerged out of non-existence.

Each marble uniquely crystallized.

Each has its own distinctive appearance.

Granted each marble may look like the others at first glance but each marble is unique and can be identified as its own self upon close examination.

So once again, where is time found to exist in the graphic?

Time is found as an element of the process of development of each marble. Thus:

Daniel J Shepard
Channel

The surroundings 'within' which the marble is immersed is void time in terms of the marbles very existence.

Does time affect the existence of the marble?

Physically time affects the physical existence of the marble but time does not affect the fact that the marble was what it was.

Time only affected what the marble became on its way to being what it is.

Thus time remains a factor found 'within' the marble but time does not affect the abstractual existence of the marble.

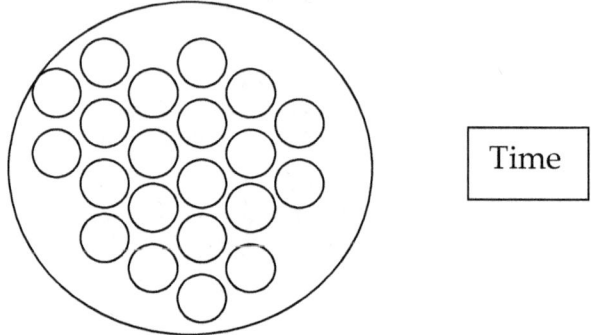

Now lets apply the concept to a unit of knowing.

Panentheism
Addressing
Anthropocentrism

24. A unit of knowing is not a marble

Units of knowing are not marbles but units of knowing do have some similarities to marbles.

Each unit of knowing was formed in its own unique manner through becoming, as was the marble.

Each unit of knowing is unique because it existed through time and had its own unique summation of input as an entity of knowing.

Each unit of knowing may have a similar appearance at first glance but upon closer examination, each unit of knowing is found to be unique.

As such, the bag of marbles might better be represented as stones in a Kaleidoscope, which adds uniquely to the potential of the whole and what it is the whole becomes.

Now using the word 'becomes' may lead one to believe that time is a factor found to surround the marbles but that is not the case.

The bag of marbles is simply a bag of marbles and has the potential to have an overall appearance of but one summation if the bag contains but one marble.

As the number of marbles increases beyond the total of one and becomes two, then three, then four, etc. so to grows the potential summation alternatives of the whole.

This growth is not geometric but rather exponential.

The point being made here, however, is not that the summation of potentiality grows but that time is not a factor of the potential but rather time is the essential element of the growth of the sub-unit emergence itself rather than an element of the whole as the whole.

So, what of units of knowing not being marbles?

Let's reexamine the bag of marbles and apply uniqueness to the marbles in terms of apparent knowing.

As such, the bag, for convenience purposed, becomes a rectangular window and the marbles become irregularly shaped objects representing 'uniqueness'.

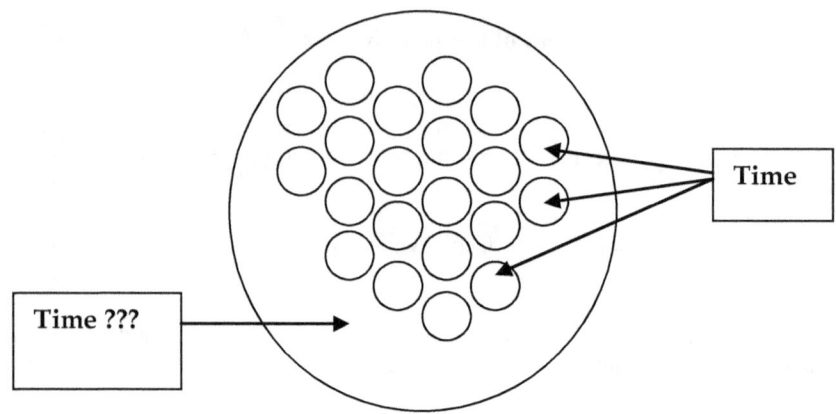

Becomes:

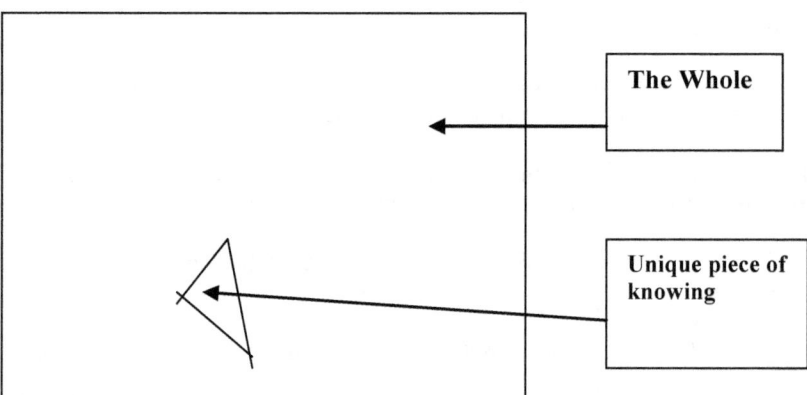

The triangle can of course rotate in any direction and as such the appearance of the whole changes.

To simplify the process we will work only in terms of two-dimensional space.

Panentheism
Addressing
Anthropocentrism

Adding another unique piece of knowing:

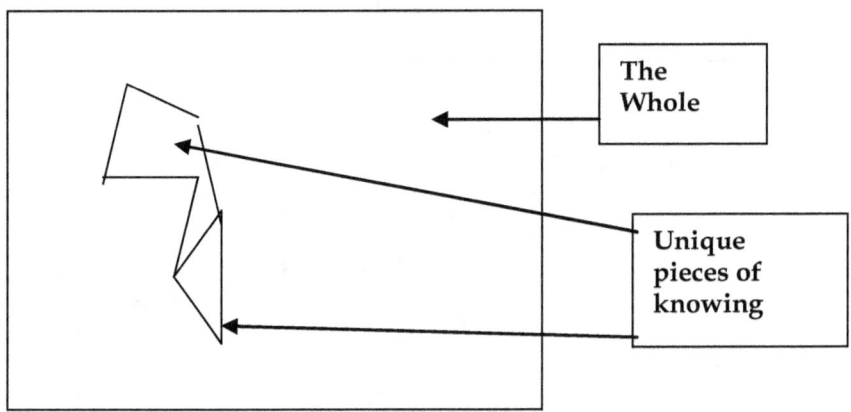

Add another unique piece of knowing:

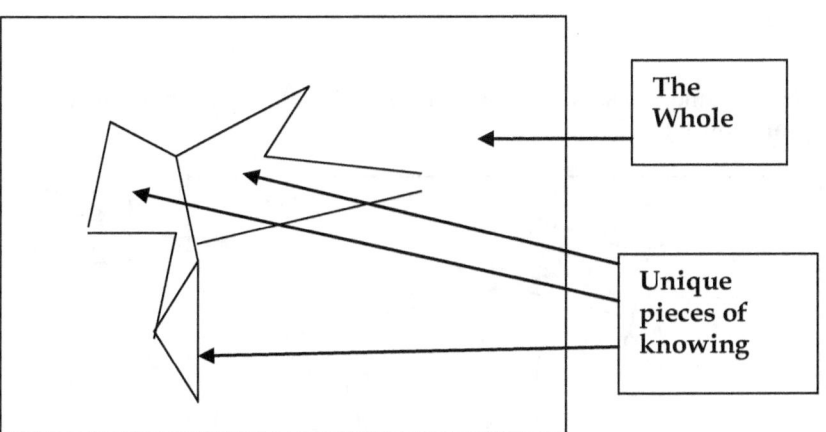

And so it is the picture grows.

Keep in mind that the picture has the potential to shift just as it does in a Kaleidoscope.

However, where do the units form?

They form through time. As such, time must have a location within which to exist.

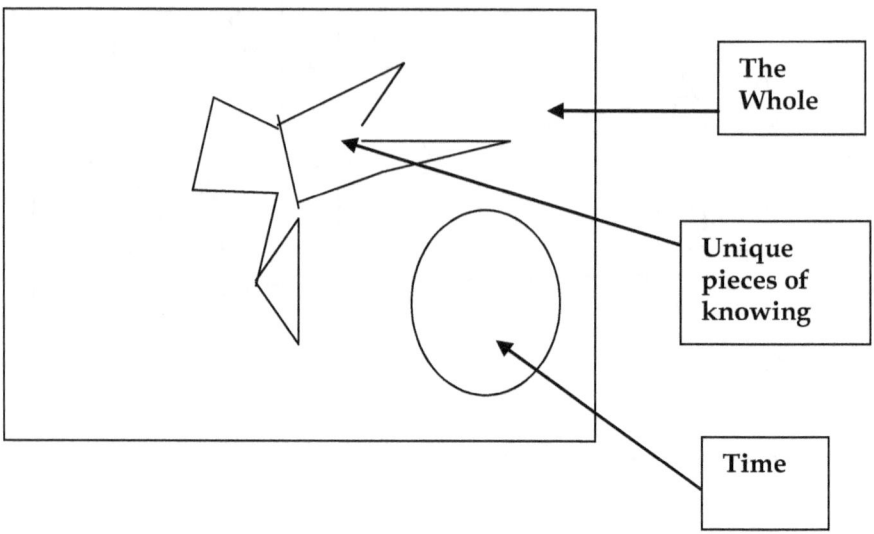

Within time unique pieces of knowing emerge from being virgin knowing to being a unit of knowing:

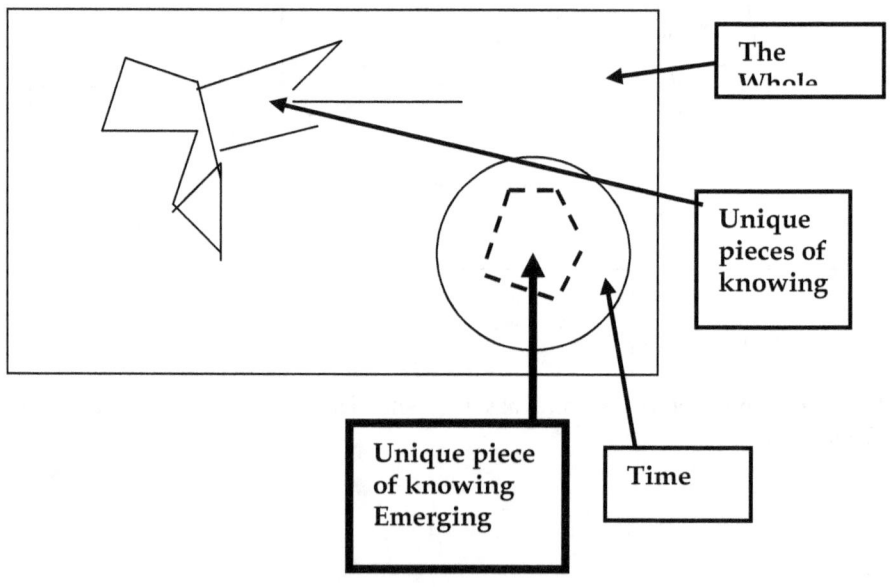

Panentheism
Addressing
Anthropocentrism

It is reaching the point where we must leave Copernicus and move on with other paradoxes.

As such, the final question becomes: What does this have to do with Centrist systems and Copernicus?

Centrist systems have a center, a central point from which unique pieces of knowing expand upon themselves, central points from which unique pieces of knowing emerge.

It is the 'unique unit of knowing', with its own unique experiencing, with its own unique knowledge which emerges from a Centrist system, emerges from 'the', 'a' physical universe and 'enters' the existence 'beyond' the physical, 'beyond the universal fabric of space and time.

The universes of which we have familiarity, which forms our particular type of unit, is one whose very fabric of apparent universality is time and space/distance which itself may be an innate characteristic of matter and energy.[xii]

Whether or not it is time and space/distance that are the innate characteristic of matter and energy or matter and energy that are the innate characteristic of time and space/distance is not the point of this tractate.

The point is: Centrist systems have a center, a central point from which unique units of knowing expand beyond and upon themselves, central points from which unique units of knowing emerge.

It is the 'unique units of knowing', which through their own unique experiencing, emerge from a Centrist system, 'the', 'a' physical universe and become a part of the non-centrist system.

This statement can be rephrased as:

Centrist systems have a center, a central point from which unique units of knowledge expand beyond and upon themselves, central points from which unique units of knowledge emerge.

It is the 'unique unit of knowledge', which through its own unique formation, knowledge itself emerges from a Centrist system of knowledge formation, from 'the', 'a' physical universe and becomes a part of the non-centrist system of total knowledge.

If we return to our marble analogy, we can begin to grasp a clearer image of what is being said:

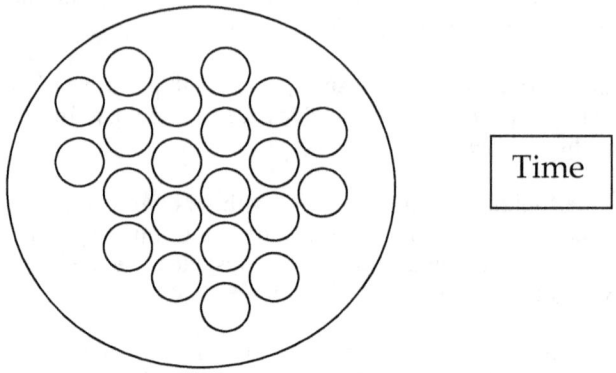

This becomes:

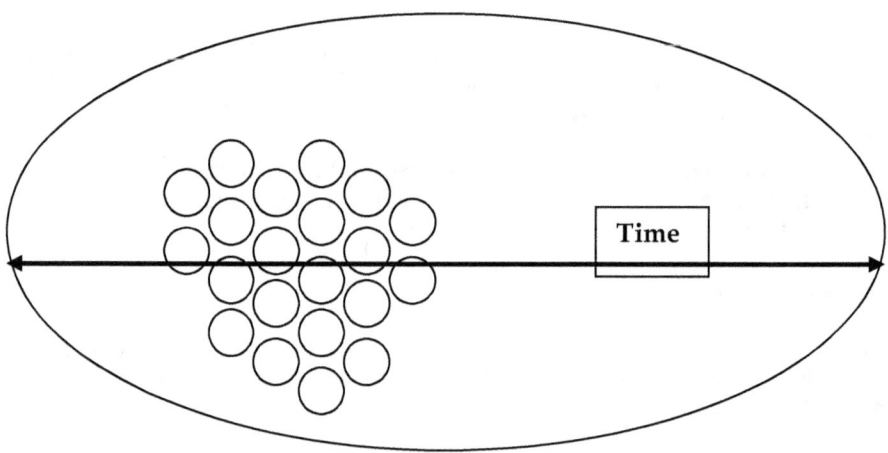

Panentheism
Addressing
Anthropocentrism

Which becomes:

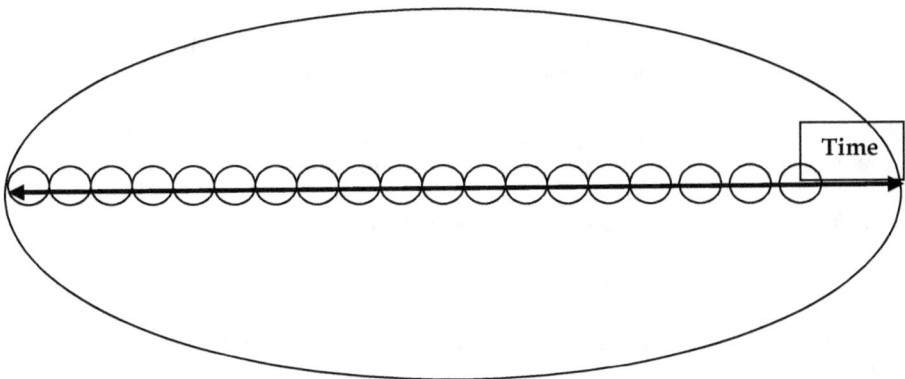

We can now see how it is that time becomes a part of each unit entity of knowing which travels through our universe of centrist systems

This however is not the end of the analogy for we have twenty-one marbles. The center is what is called the origin, is called the present, is defined as: What is:

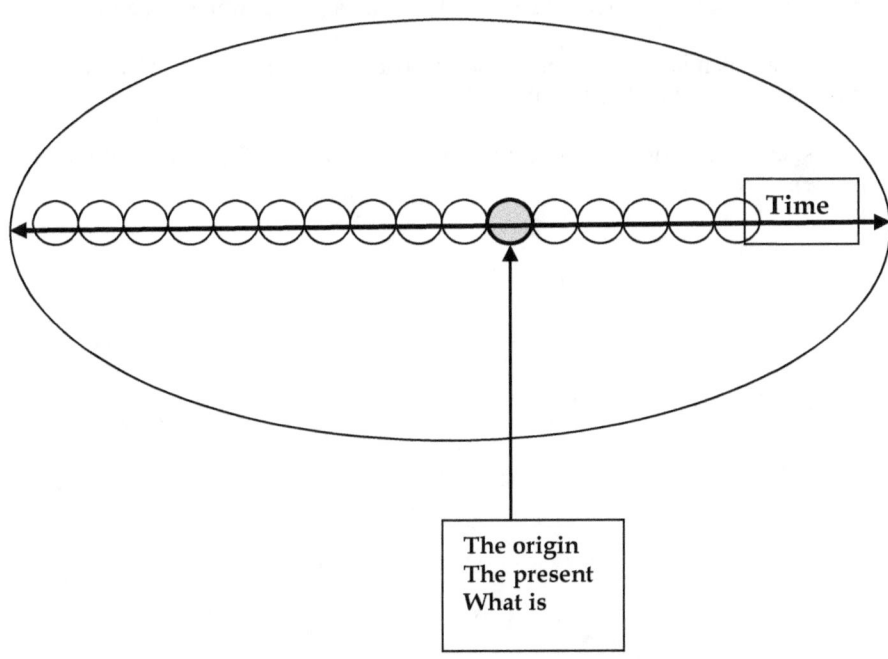

The left of the center units, the left of the present are units, which have already "become". We refer to such a concept as the past or 'what was'.

The right of the center unit, the right of the present are units, which have yet to "become". We refer to such a concept as the future or 'what could be'.

The concept of the entities to the right of 'what is' being 'what could be' versus 'what will be' is a very important aspect of the future events.

'What could be' implies free will of the individual.

'What will be' implies determinism.

It would be interesting running this train of thought under the premise of 'what will be'.

That scenario, however, belongs with Boethius and we are not going back there.

We must stay on track of following history and where it has led us, is leading us, and 'could' lead us.

Let it be noted, however, that in the case of the 'what will be' scenario, the marbles to the right of center are all pre-filled with a predetermined pattern.

As such the potentiality of kaleidoscope of patterns is limited to 'what is' rather than being unlimited by 'what could be'.

Within such a scenario, infinite finites supercede finite infinites (see Volume 7: Aristotle).

Panentheism
Addressing
Anthropocentrism

Returning to the task at hand, we then obtain:

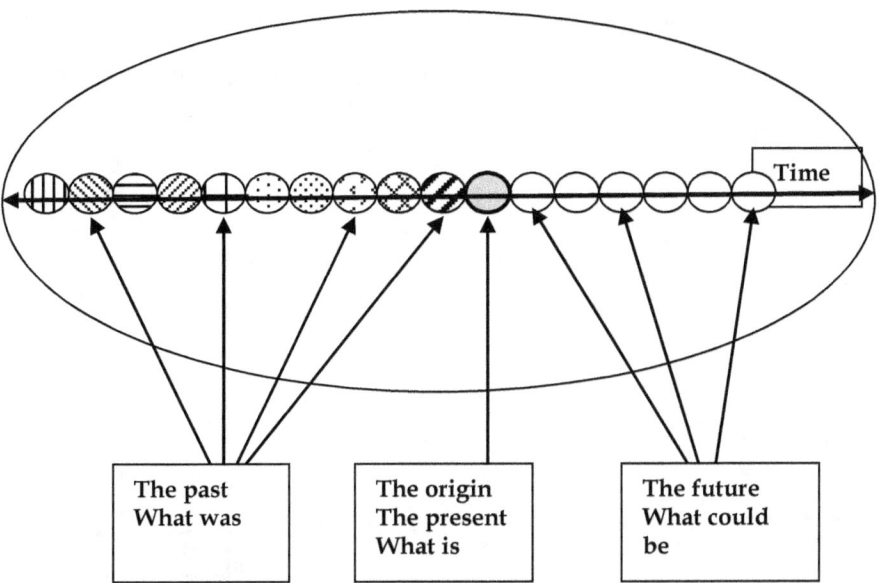

Now of course we recognize the left side of 'what is' as not only the stones of the Kaleidoscope existing within the void of time and space/distance yet 'containing' perceptions of time and space/distance.

We can, therefore redo our graphic and in doing so we find we have:

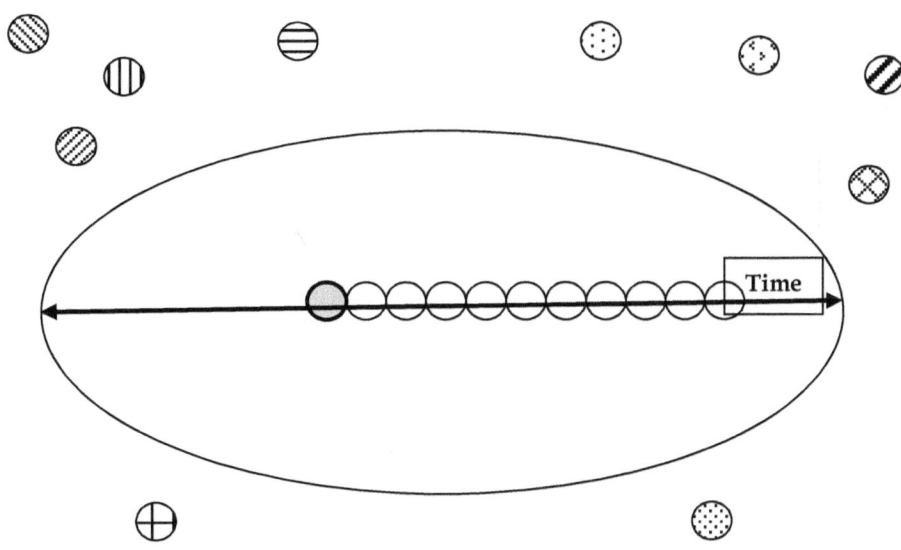

Daniel J Shepard
Channel

As the marbles roll through space and time, they become unique one from another.

This is a two dimensional depiction. One must not lose track of the concept that even in a two-dimensional depiction time has many degrees of latitude regarding 'where' it is headed:

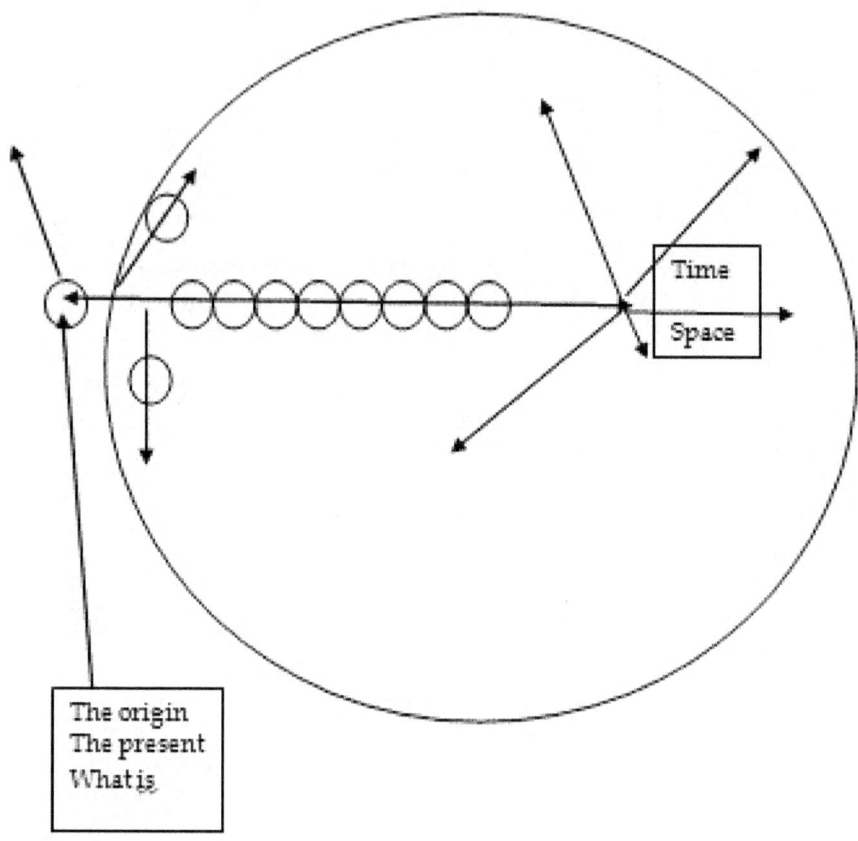

Panentheism
Addressing
Anthropocentrism

In essence, we have:

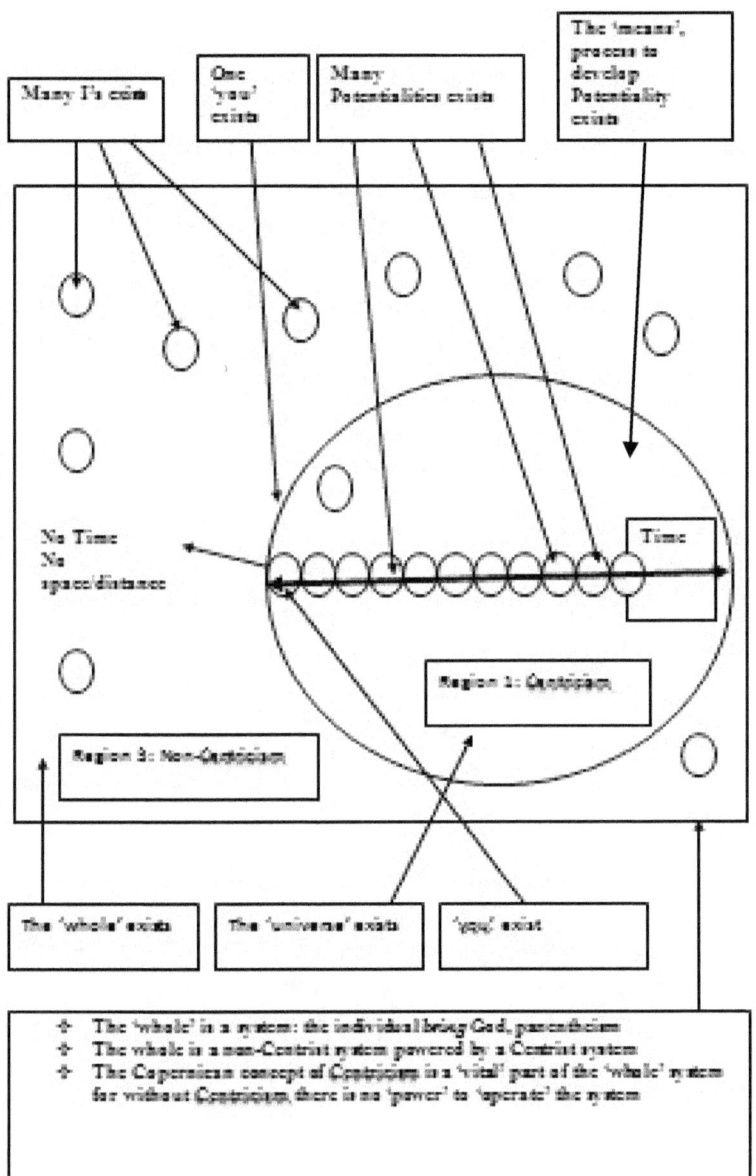

- The 'whole' is a system: the individual being God, panentheism
- The whole is a non-Centrist system powered by a Centrist system
- The Copernican concept of Centricism is a 'vital' part of the 'whole' system for without Centricism, there is no 'power' to 'operate' the system

Daniel J Shepard

Channel

With the aid of panentheism we now understand that Copernicus is a vital link in moving our perceptual understanding forward regarding the 'system' being filled with Centrism into that of being 'the' system filled with Centrism and non-Centrism.

As such, Centrism and non-Centrism, through the help of a panentheistic model and with the help of Copernicus, now are understood to have a location within which they can be found.

And now, the understanding regarding the role of both Centrism and non-Centrism as well as the understanding regarding the interrelationship between Centrism and non-Centrism no longer remain in a state of confusion.

Even more interestingly, the existence of such an interrelationship is, with the help of panentheism, is not only recognized as a significant aspect of the 'larger' system but it is now understood how centrism and non-centrism interact one with the other.

Panentheism
Addressing
Anthropocentrism

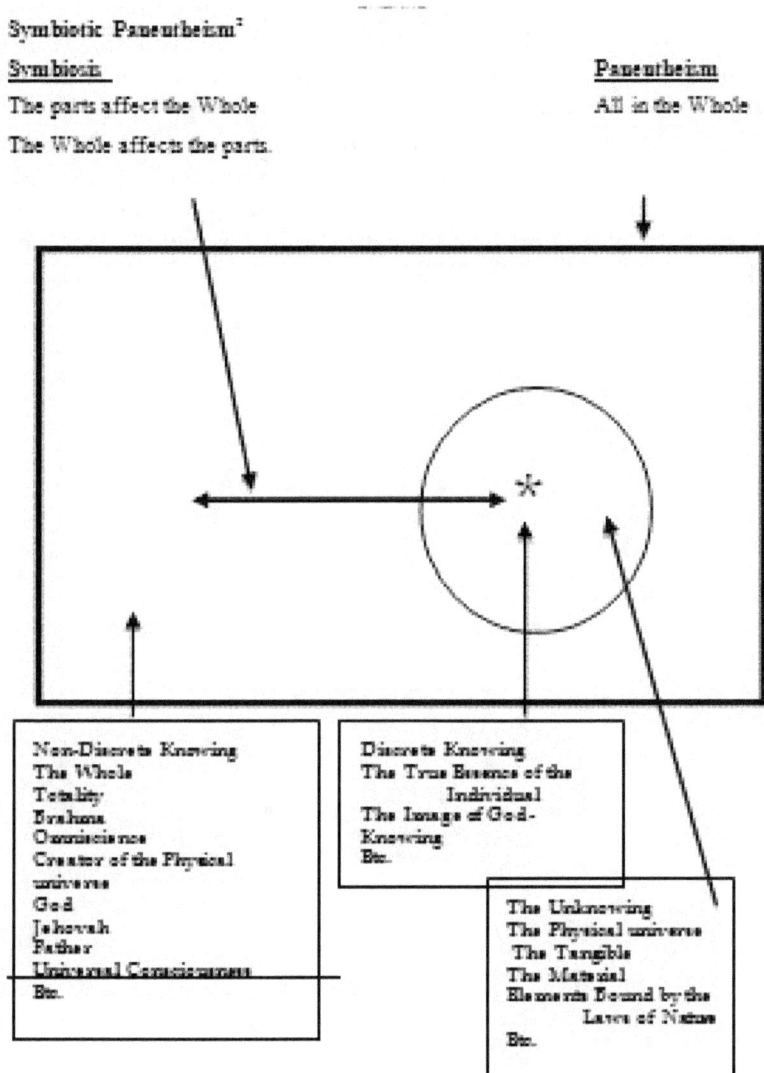

2000 CE
Understanding Reality Evolving

Daniel J Shepard
Channel

[1] Why add the adjective 'Symbitotic' to the noun Panentheism? There are, as with all nouns, many subgroups of the noun. In the case of panentheism there are many types of panentheism. Within the works produced by the author it is panentheism which provides the answers to the the third question: Why? Why does the physical universe exist? Why do we, you and I, exist? Why did Universal Consciousness create Discrete Consciousness? Why was nothingness created? Why have we been unable to resolve age old philosophical, religious and scientific paradoxes and puzzles? Etc.

Panentheism
Addressing
Anthropocentrism

About the author and the work

About the author and the work

In 1951, after contact with universal consciousness, I began examining the metaphysics of reality. Being quite young at the time, I did not understand the concept of universal consciousness. Being too young to understand the concept of the universal consciousness did not make the experience any less an experience.

My experiences with the spiritual side of reality placed me in a position of eventually having to move on and live in the world of materialism or to accept the responsibility to reveal what it was I was being given. The decision did not come easily and the decision could have easily gone either way.

There was a real fear I had to confront. Was I in touch with 'evil' or 'good' and just what would be the consequences in an after life should an after life exist. Of course if no after life existed, there would be no consequences to face, there would simply be a peaceful sleep.

I was not afraid of a lack of an after life, for there being or not being an after life was beyond my control. Likewise, if there was no free will, whatever decision I made was not only meant to be but also destined to be.

The thing that concerned me was free will. Did free will exist or not? If free will did exist I would be responsible for my actions. As such I had a decision to make and the decision was one that could have a major impact upon my soul, upon my 'eternal', timeless, existence.

My twin sister had the same concerns and weeping, begged me to reconsider what I was doing. She believed the risk was far too great to make the commitment of revealing what it was I was given. Her greatest fear was that she would never see me again after death.

Internally I seriously debated that very point. To make light of the issue, however, I would say to her that if she went 'up' and I went 'down', we could still talk to each other through the street drains that separated the two worlds.

As much as I tried to add levity to what it was I was about to do, the decision was by no mean taken lightly by me. One does not joke about one's soul.

Daniel J Shepard

Channel

Science/observations, philosophy/rational dialectics, ancient wisdom/universal principles of faiths and mathematics/universal language clearly reinforce what it is that had been and was continuing to be revealed to me, namely: There is, beyond all reasonable doubt, an eternal/timeless side to reality and our/humanity's true substance and true essence played a unique role in the ethereal side of reality.

My science, mathematics, rational and spiritual exposures all appeared to be in sync. This apparent synchronization combined with the observation that the model of reality being unveiled produced nothing but long term 'good', convinced me to move forward with project.

The model being revealed clearly demonstrated that, we, you and I, humanity has a responsibility, no, not just a responsibility but a great responsibility to what it is the ethereal side of reality is in the process of becoming. We, you and I, humanity are partially responsible for what it is we will eventually find ourselves immersed within when we step into a timeless existence, step into eternity. This is no mundane matter.

The responsibility cannot be taken seriously, however, if we do not understand the responsibility. If we do not understand the repercussion of our actions. Acting upon our intuitive sense of faith alone, acting upon faith void reason, does not appear to be enough for large segments of humanity in this day and age, thus the need for these books.

I do not take credit for the ideas presented in these works for they did not come from me but rather through me, came from elsewhere.

I am not saying the words themselves are divine in nature for clearly they are not. In fact I am truly embarrassed by my amateurish attempts at writing. I did not end up taking science and mathematics classes because I was gifted in those fields. I took science and mathematics classes because I detested writing and as such social studies and English classes were classes I did my utmost to avoid, yet here I am swallowed up in the written word I so consciously attempted to avoid. Ah, the irony of it all.

So it is the process of recording what it was I was given began in 1995 with the first publication in 1996 titled: You and I Together – Have a purpose in reality. This first work was followed by a continual stream of works intended to provide overwhelming evidence regarding the validity of the model describing the whole of reality and our, your, my function within such an existence.

Before listing the works presently available to the public, I would like to add a few words regarding the political correctness of today's perception of relativism since this work all demonstrates absolutism is not only reasonable but also rational and in fact trumps relativism.

Panentheism
Addressing
Anthropocentrism

Some would say one person's philosophical perceptions are no more valid than another's. Such a perception runs rampant in today's society. The perception that everyone's understanding of reality is no less valid than another's is what is called relativism.

To suggest everything is relative is both illogical and irrational. Rational relativist would never embrace the idea that anyone's art is as good as Van Gough's, Rembrandt's, Picasso's or Dali's. Nor would the logical relativist agree the physics of a freshman physics student is as good as that of Einstein's, Hawking's, Hubble's or Newton's. Show me a relativist who would agreeably undergo surgery performed by a high school biology student when a trained surgeon was available and I'll show you a relativist who is not rational.

In our present age, these same relativists would unquestioningly defend the position that reality is what each person personally perceives reality to be. The relativist holds there is no such thing as absolute truth and therefore the relativist argues good and evil are simply constructs of cultures and societies built by humans themselves and change as culture and social structures change.

The model of reality demonstrated by the project does not suggest relativism is completely erroneous rather the works demonstrate relative truth to be secondary to absolute truth. Issues may appear to be relative to the individual but the individual lies within an absolute existence, lies within an absolute reality and it is this absolute existence from which absolute morals and absolute ethics emerge. We, each of us, is capable of understanding this absolute existence. We as a species are capable of using a model of the whole of reality as the foundation upon which we base both our decision to act and our decision to not act.

The understanding of reality does not replace religion. The understanding of reality reinforces religions, which clearly demonstrate that we are all children of God/Jehovah/Brahma and as such each and every one of us should be treated as we would treat God should God stand before us.

Daniel J Shepard
Channel

Technically this 3rd millennial philosophical evolutionary development becomes:

<u>Absolutism exists and</u>
<u>Relativism is non-existent</u>

Thesis:

<u>Absolutism is the universal fabric of the 'exterior' of the individual, the whole within which the individual exists</u>

        Leading to:

Synthesis: And

<u>Relativism is the universal fabric of the 'interior' of the individual, the experiencing of which the individual knows</u>

Antithesis:

<u>Relativism exists and</u>
<u>Absolutism is non-existent</u>

The result: Absolutism trumps relativism, altruistic hedonism supersedes physical hedonism

The books, blogs, power point presentations and audios provide unambiguous answers to three questions:
    1. Where you are in terms of the whole of reality
    2. What you are in terms of the whole of reality
    3. Why you exist in terms of the whole of reality

So much for a few words.

Panentheism
Addressing
Anthropocentrism

[i] Aristotle initiated the elementary form of Cartesianism, which might better be termed 'static Cartesianism'. Although some discussion of 'static' Cartesianism will occur in this tractate, a detailed explanation of what the concept means to metaphysics can be found in Volume 7: The Error of Aristotle.

[ii] **Question**: This paragraph suggests philosophy, science, and religion were equals. Before Copernicus, philosophy did enjoy an independent status apart from religion. However, by the medieval period, philosophy was used to support – and not compete with – religious views. It wasn't until the emergence of science, as an independent area of study, that philosophy once again regained its independence. At that point – philosophy, religion, and science – all became independent of each other. With this in mind could you explain what you are attempting to imply? **Answer**: We have at our disposal three means of developing perceptions. We have science/observation, religion/belief, and philosophy/reason. Each of the three, throughout the unfolding of time, wanes and ebbs in terms of its 'apparent' significance one to the other. But in truth each is, uniformly through time, equal to the other in significance and any universally stable perception we develop has no choice but to be confirmed by the three equally. We will never, as individuals or as species, accept the validity of a metaphysical model if it is a model: We find unsupportable by either direct or indirect observation, interpolation or extrapolation, and induction or deduction. We find to be unbelievable. We find to be unreasonable.

[iii] **Question**: Because the foundations of political philosophy were being established during this period, I don't perceive the process of the "Why" as being a random and unimportant event. I agree that, during this period, we "settled" for different models of cooperation and agreement, rather than a larger metaphysical understanding. However, these earlier "models" did help establish an atmosphere whereby later philosophers could propose more advanced "models'. As such, can you clarify what you mean by 'random' and 'unimportant' events? **Answer**: Metaphysically speaking, no 'foundation of reason' was being laid down to rationalize the concept regarding 'tolerance and respect due the individual'. The understanding of 'a' metaphysical model from which a natural emergence of such a perspective would occur had not been laid out for students of philosophy to examine. Granted the emergence of political philosophy was just emerging but that is not the point. Political philosophy is no more a foundation of reason than is religion or science. The foundation for the concept regarding 'tolerance and respect due the individual' can be found in science/observation, religion/faith, and philosophy/reason. Philosophy/reason, itself has 'a' foundation and that foundation is the most basic, the most primitive, the most primal of foundations. This most primal, most basic of foundation is the understanding of 'a' metaphysical system which explains the very fundamental dynamics existing between ourselves as abstractually knowing individuals and what lies beyond the physical itself.

[iv] **Question**: Can you clarify? **Answer**: Up to and through Copernicus, the West, for the most part, 'believed' time and space/distance were aspects of the physical. Such perceptions dominated not only scientific thought, but also religious and

philosophically thought. With the advent of Kant, however, such philosophical perceptions underwent the same type of traumatic inversion as occurred to science and its concept of Centrism with the advent of Copernicus.

[v] Oxford Concise Science Dictionary, 1996

[vi] **Question**: Can you clarify? **Answer**: Up to and through Copernicus, the West, for the most part, 'believed' time and space/distance were aspects of the physical. Such perceptions dominated not only scientific thought, but also religious and philosophically thought. With the advent of Kant, however, such philosophical perceptions underwent the same type of traumatic inversion as occurred to science and its concept of Centrism with the advent of Copernicus.

[vii] **Question**: "…lack of time and distance"? Can you explain? **Answer**: All forms of physical existence are tied to time and space. Physical existences find their very existence defined by four coordinates: the three dimension of space and the dimension of time. Without space and time physical objects could not exist, as we know them to be, physical. Abstract concepts on the other hand are not dependent upon the physical quality of neither space nor the physical quality of time. Granted some abstractual concepts are dependent upon an abstractual understanding of space and time but they are, however, not dependent upon the very existence of the physical qualities of space and time themselves.

[viii] **Question**: Can you clarify? **Answer**: No, other than to say that omniscience by definition leads to the summation of knowing, summation of motivation of action. As such, Gandhi and Hitler, both of whom impacted human awareness through two opposing points of view, became a part of total knowing. The result, the summation of knowing, omniscience, incorporated the aspects of both men as well as the horrendous number of ripples both men initiated which in turn became, are still becoming, will continue to become part of the summation of knowing and thus mold the very personality of the summation of knowing itself.

[ix] Genesis 32:30, 8:2

[x] **Boethius: The Consolation of Philosophy, Book V, section IV.**

[xi] **Question**: Can you expand on this statement? **Answer**: Husserl would suggest all things can be stripped away from reality until only one primary concept remains. Such a process would leave the most fundamental of foundations. Such a foundation would be termed 'the Archimedean Point' from which all else emerges.

[xii] **Question**: Can you clarify? **Answer**: Up to and through Copernicus, the West, for the most part, 'believed' time and space/distance were aspects of the physical. Such perceptions dominated not only scientific thought, but also religious and philosophically thought. With the advent of Kant, however, such philosophical perceptions underwent the same type of traumatic inversion as occurred to science and its concept of Centrism with the advent of Copernicus.

www.ingramcontent.com/pod-product-compliance
Lightning Source LLC
Chambersburg PA
CBHW051700170526
45167CB00002B/478